ROYAL BOTANIC GARDENS KEW

ROYAL
BOTANIC
GARDENS
KEW

WORLD HERITAGE SITE

Souvenir Guide

Contents

Welcome to Kew

In 2009 we celebrate 250 years as a botanical garden, providing inspiration and sharing knowledge about the world's plant life. Kew today is a global centre for research on plants and fungi, as well as a site of World Heritage, rich in history and with the Earth's largest botanical and horticultural collections under its care. Increasingly, we are reaching out across the world in partnerships with other organisations aiming to conserve plant life and chart a way forward that ensures a sustainable future for plants and people.

I hope you enjoy your time at Kew, and come again to see its richness and diversity through the seasons. Above all, I hope you find inspiration to help care for the world's plants and fungi – the basis of all life – and discover ideas that help you minimise your impact on the planet's increasingly threatened environmental resources in a time of rapid climate change.

Professor Stephen Hopper
Director, Royal Botanic Gardens, Kew

About Kew Gardens and this book

The Gardens cover an area of over 121 hectares (300 acres) on the south bank of the Thames in south-west London. The vast number and variety of plants means the nature of Kew Gardens changes with the seasons. Out in the grounds and inside the plant houses, thousands of specimens progress through their annual cycles, flowering, fruiting, growing, resting. All year long, there are plants to be seen at their glorious best.

But Kew is much more than one of the world's finest showpiece gardens. It's an internationally respected centre of scientific excellence, identifying and classifying plants; researching their structure, chemistry and genetics; collecting and conserving endangered species; maintaining reference collections and sharing all this knowledge with interested parties throughout the world.

This book tells you what there is to see, where it is, and how to get there. On the next page, a map gives basic information about the Gardens, the public facilities, the main plant houses and plantings.

Kew is a very seasonal pleasure, so we take you through the seasons, with maps and a choice of walks to where some of our experts' favourite plants are at their best. Then come separate sections for the major buildings and garden areas, talking about their key features and a few more details, combining a little history with a lot of interesting facts. Finally, there's Kew's history and its role today - more important now than ever in its lifetime of almost 250 years.

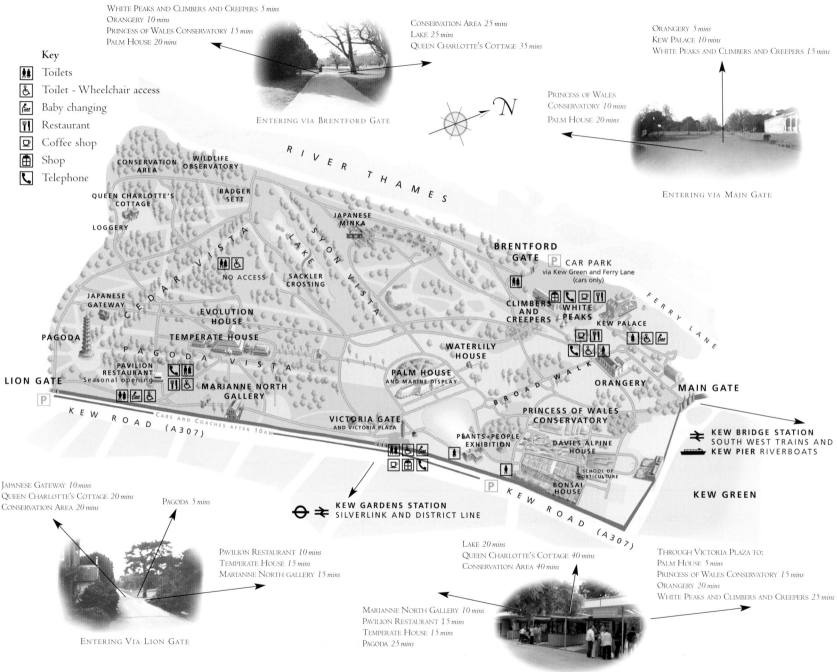

Key
- 👫 Toilets
- ♿ Toilet - Wheelchair access
- 🚼 Baby changing
- 🍴 Restaurant
- ☕ Coffee shop
- 🛍 Shop
- ☎ Telephone

WHITE PEAKS AND CLIMBERS AND CREEPERS *5 mins*
ORANGERY *10 mins*
PRINCESS OF WALES CONSERVATORY *15 mins*
PALM HOUSE *20 mins*

CONSERVATION AREA *25 mins*
LAKE *25 mins*
QUEEN CHARLOTTE'S COTTAGE *35 mins*

ENTERING VIA BRENTFORD GATE

ORANGERY *5 mins*
KEW PALACE *10 mins*
WHITE PEAKS AND CLIMBERS AND CREEPERS *15 mins*

PRINCESS OF WALES CONSERVATORY *10 mins*
PALM HOUSE *20 mins*

ENTERING VIA MAIN GATE

N

RIVER THAMES

CONSERVATION AREA
WILDLIFE OBSERVATORY
QUEEN CHARLOTTE'S COTTAGE
BADGER SETT
LOGGERY
JAPANESE MINKA
CEDAR VISTA
LAKE
SYON VISTA
NO ACCESS
SACKLER CROSSING
JAPANESE GATEWAY
EVOLUTION HOUSE
TEMPERATE HOUSE
PAGODA VISTA
PAGODA
PAVILION RESTAURANT *Seasonal opening*
MARIANNE NORTH GALLERY
WATERLILY HOUSE
PALM HOUSE AND MARINE DISPLAY
BROAD WALK

BRENTFORD GATE
P CAR PARK *via Kew Green and Ferry Lane (cars only)*
CLIMBERS AND CREEPERS
WHITE PEAKS
KEW PALACE
FERRY LANE
ORANGERY
PRINCESS OF WALES CONSERVATORY
DAVIES ALPINE HOUSE
PLANTS+PEOPLE EXHIBITION
SCHOOL OF HORTICULTURE
BONSAI HOUSE

LION GATE
P
KEW ROAD (A307)
Cars and Coaches after 10am

VICTORIA GATE AND VICTORIA PLAZA

MAIN GATE

🚉 KEW BRIDGE STATION
SOUTH WEST TRAINS AND
🚢 KEW PIER RIVERBOATS

KEW GREEN

P KEW ROAD (A307)

KEW GARDENS STATION
SILVERLINK AND DISTRICT LINE

JAPANESE GATEWAY *10 mins*
QUEEN CHARLOTTE'S COTTAGE *20 mins*
CONSERVATION AREA *20 mins*

PAGODA *5 mins*

PAVILION RESTAURANT *10 mins*
TEMPERATE HOUSE *15 mins*
MARIANNE NORTH GALLERY *15 mins*

ENTERING VIA LION GATE

LAKE *20 mins*
QUEEN CHARLOTTE'S COTTAGE *40 mins*
CONSERVATION AREA *40 mins*

MARIANNE NORTH GALLERY *10 mins*
PAVILION RESTAURANT *15 mins*
TEMPERATE HOUSE *15 mins*
PAGODA *25 mins*

ENTERING VIA VICTORIA GATE

THROUGH VICTORIA PLAZA TO:
PALM HOUSE *5 mins*
PRINCESS OF WALES CONSERVATORY *15 mins*
ORANGERY *20 mins*
WHITE PEAKS AND CLIMBERS AND CREEPERS *25 mins*

What to see at Kew

WHATEVER YOUR INTEREST, FROM PLANTS BOTH HUMBLE AND EXOTIC; THROUGH WONDERFUL LISTED BUILDINGS FROM A BYGONE AGE; TO FUN INTERACTIVE LEARNING FOR YOUNG ONES; IF IT'S AT KEW, IT'S MORE THAN LIKELY TO BE AMONG THE BEST AROUND.

TO HELP YOU MAKE THE MOST OF YOUR VISIT, THE MAP OPPOSITE SHOWS WHAT CAN BE FOUND AND WHERE IN THE GARDENS YOU CAN FIND IT. TRY A SEASONAL WALK, CREATED BY KEW'S EXPERTS, TO SHOW YOU WHAT'S LOOKING ITS BEST AT DIFFERENT TIMES OF THE YEAR.

Kew's iconic buildings
The Palm House and the Temperate House are known all over the world as symbols of Kew. Add no fewer than 37 other listed buildings, the award-winning Princess of Wales Conservatory and the Davies Alpine House, and Kew's selection as a World Heritage Site is easily understood.
See pp 22-47 for Kew's main glasshouses; pp 66-75 for decorative buildings.

Exotic plantings
Orchids and carnivorous plants in the Princess of Wales Conservatory; Kew's oldest pot plant in the Palm House; a cycad, possibly the last of its species in the world, in the Temperate House; a thriving example of the 'pinosaur' Wollemi Pine near the Orangery — Kew's beautiful, economically important, and rare plants can be found throughout the Gardens.

Spring flowerings
Literally millions of spring bulbs; a bluebell wood; walks through trees in blossom; a wonderful show of rhododendrons and azaleas; spring bedding displays - it's now that Kew comes to life.
See pp 6-9 for Kew's own view of spring and two delightful walks.

Summer beauties
Outside in the Gardens and inside the glasshouses, the sights and scents of summer when Kew is in full bloom crowd the visitors' senses. Enjoy flowering shrubs and the superb bedding display outside the Palm House.
See pp 10-13 where Kew's experts talk about summer and offer two walks to enjoy.

Autumn glories
With colour coming to leaves, bark and ripening berries; with ornamental grasses at their best; and with Europe's finest collection of mature hollies to see; some visitors enjoy autumn at Kew most of all.
See pp 14-17 to see why and perhaps to follow our experts' walks.

Winter interest
There's more to see in winter than many people think. Christmas roses, snowdrops and winter-flowering clematis; wintersweet and witch hazels with their heady scents and the welcome of tropic warmth in the glasshouses. Kew's Christmas events usher in the real festive spirit.
For more interest and two invigorating walks, see pp 18-21.

Kew for kids
Kew plants ideas about the environment in young inquisitive minds. There's an ongoing programme with schools; while families can have fun learning more with interactive displays in Plants+People (*see p 75*); at the Wildlife Observatory, the Stag Beetle Loggery and walk-in Badger Sett (*see p 63*) and, especially for young ones, in Climbers and Creepers (*see p 88*).

Art at Kew
Botanical art is always impressive, from the scientifically accurate to looser, more impressionistic studies. See the permanent exhibition of paintings in the Marianne North Gallery; while there is more contemporary work in the Kew Gardens Gallery; and regularly changing shows of artworks and photography around the Gardens (*see pp 76-77*).

NEAR NEIGHBOURS

Rhododendrons and azaleas are the same plant to botanists, but not to gardeners. Here in the Gardens they are near neighbours and make a fine display.

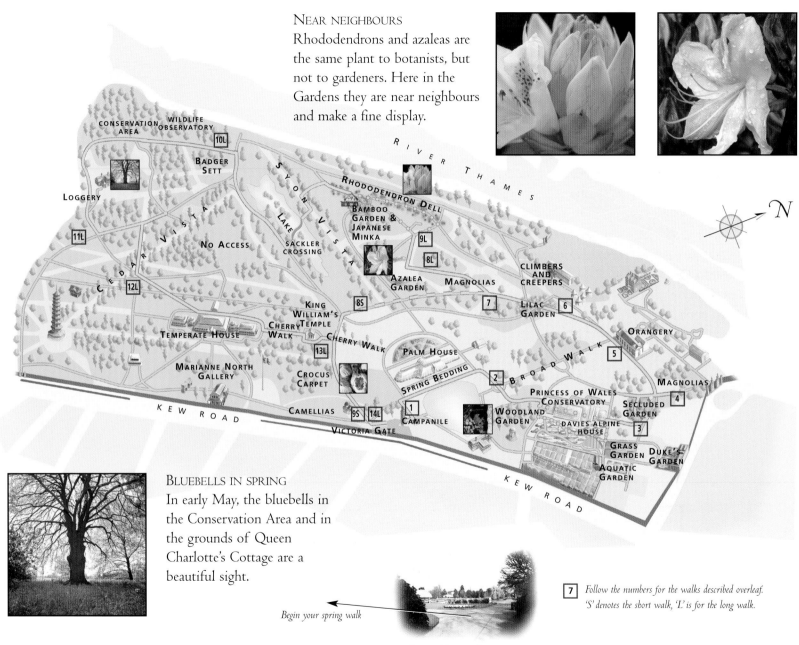

CONSERVATION AREA

WILDLIFE OBSERVATORY

10L

BADGER SETT

LOGGERY

11L

12L

CEDAR VISTA

NO ACCESS

LAKE

SACKLER CROSSING

SYON VISTA

RHODODENDRON DELL

BAMBOO GARDEN & JAPANESE MINKA

9L

8L

AZALEA GARDEN

MAGNOLIAS

CLIMBERS AND CREEPERS

RIVER THAMES

8S

KING WILLIAM'S TEMPLE

7

LILAC GARDEN

6

ORANGERY

TEMPERATE HOUSE

CHERRY WALK

CHERRY WALK

13L

PALM HOUSE

SPRING BEDDING

BROAD WALK

5

MARIANNE NORTH GALLERY

CROCUS CARPET

2

PRINCESS OF WALES CONSERVATORY

MAGNOLIAS

SECLUDED GARDEN

4

CAMELLIAS

9S

14L

1

CAMPANILE

WOODLAND GARDEN

DAVIES ALPINE HOUSE

3

VICTORIA GATE

KEW ROAD

GRASS GARDEN

DUKE'S GARDEN

AQUATIC GARDEN

KEW ROAD

N

BLUEBELLS IN SPRING

In early May, the bluebells in the Conservation Area and in the grounds of Queen Charlotte's Cottage are a beautiful sight.

Begin your spring walk

LEAVING VICTORIA PLAZA

7 *Follow the numbers for the walks described overleaf. 'S' denotes the short walk, 'L' is for the long walk.*

Spring at Kew

Seasons and flowering times, especially in spring, can vary by up to three or more weeks, so the periods below are only approximate. For up-to-date details of flowering, please call 020 8332 5655 or visit the website at www.kew.org

February to May sees the full spread of spring from fresh awakenings to bravura displays. There are crocuses en masse between the Victoria Plaza and King William's Temple and camellias on the way to the Marianne North Gallery. Daffodils and narcissi are naturalised all over the Gardens and the copious plantings of daffodils either side of the Broad Walk are a fine sight.

The Cherry Walk, with 300,000 scillas planted beneath its trees, can be stunning, along with the total of more than 900,000 wild daffodils, crocuses and snake's head fritillaries planted between the Lilac Garden and the Magnolia Collection. See the sea of *Chionodoxa* on the lawn near White Peaks and find out how they got the name 'Glory of the snow'. The spring bedding outside the Palm House is a definite must-see, and elsewhere, crab apple trees and wisterias are in flower. Make sure you visit the Queen's Garden as well, where the laburnum arch will start to flower in late spring.

There's much to see in the Conservation Area with carpets of bluebells (late April to early May), a walk-in badger sett and a skyscraper 'loggery' for protected stag beetles.

You can find the main flowering displays on this map, and to try a favourite spring walk, see the next page.

THE CROCUS CARPET
Reader's Digest sponsored the carpet of 1.6 million crocuses between King William's Temple and the Victoria Plaza on its 50th anniversary in 1987. They then funded the planting of 700,000 new corms completed in 2004. This is surely one of Kew's best-loved harbingers of spring.

BABES IN THE WOOD
The Woodland Garden comes fully to life with both emerging blooms, and the sights and sounds of wildlife busily building nests and feeding after the long winter.

Spring Walks

You can take a long or a short version of this walk. At a gentle pace, looking at plants, but not lingering, the short walk takes about one and a half to two hours; the long walk about an hour longer, starting and ending at the Victoria Plaza. You can follow the walk, which is on both paths and grass, from the description below and/or from the spring map on the previous page.

Bluebell

Magnolia

Camellia

1 Go through the Victoria Plaza, past the Campanile, and immediately turn left towards the Palm House. Turn right to walk the length of the Palm House, with the spring bedding displays on your left. There may be some waterfowl displaying on the Pond hoping to attract a mate.

2 Turn right, away from the Palm House, go straight on at the roundabout and the Woodland Garden is ahead, bursting into life. After you have sampled its delights you can either go through the Princess of Wales Conservatory to the north end, where there's an attractive seasonal display, or walk through the Rock Garden, where there's not only a superb collection of snowdrops and fine showings of other spring flowers, but also the award-winning Davies Alpine House.

3 From the Princess of Wales Conservatory north exit, go straight ahead to where you see a huge stone pine tree. Follow the brick path left into the Secluded Garden. Walk through the garden to the other side where, a little to the left, there's a wonderful display of wisteria on an old pergola.

4 As you leave the Secluded Garden head right towards the Main Gate. Just past the path to the Orangery, there's a planting of magnolias and a spectacular *Cornus* 'Ormonde', which shouldn't be missed.

5 Go to the Orangery - maybe stopping there for a little light refreshment. Look down the Broad Walk with, in February to March, a mass of daffodils on either side - Wordsworth's 'golden host' is a perfect description.

6 With the Broad Walk on your left, head up the main path to an intersection by the pretty Lilac Garden, at its best in May. Now take the path forking left towards a large cedar tree, and stroll along past wild daffodils, crocuses and snake's head fritillaries, and a magnificent display of magnolias.

short walk

long walk

7 Short walk or long walk?

Now you make your decision. For the short walk (around 15 minutes back to the Victoria Plaza), continue along the path past the magnolias. For the long walk (an extra hour, maybe more), turn right across the grass, following the Azalea Garden signs. The azaleas are an eye-opener in April and May.

Short walk:

8S Continue straight on after the magnolias, crossing one path and arriving at where five paths meet. Follow the sign to Victoria Gate and after a few paces, the splendid Cherry Walk crosses the path. Stop and look right to see King William's Temple and the Mediterranean garden and left for the Palm House. Then continue straight over this crossroads towards the Temple of Bellona, to find around two million crocuses in early spring.

9S At the path by the Kew Road wall, look right at the camellias all along the path to the Marianne North Gallery - a unique collection of flower paintings well worth a visit. Otherwise, turn left and you will arrive back at the Victoria Plaza.

Long walk:

8L Walk over the grass into the Azalea Garden. After reaching the circle of beds, turn right to join the path heading down towards the river, the Bamboo Garden and Rhododendron Dell.

9L There's a choice here - either take the path left to the Bamboo Garden, with the fascinating traditional Japanese Minka, or the one taking you through Rhododendron Dell. On the other side of these gardens take the path heading left which will take to Syon Vista. Stroll down the vista towards the Palm House and then turn right onto the elegant Sackler Crossing over the Lake. On the other side of the Lake, go to the path, turn right and head towards the Badger Sett.

10L Turn left at the signpost to Queen Charlotte's Cottage. The Badger Sett is on the left. Continue through the woods and look out for the Wildlife Observatory on the right. Continue past the Cottage with its bluebells to the dramatic Stag Beetle Loggery. Continue on this path until you reach a crossing of the paths.

11L If you have time for a diversion, turn left to see some magnificent giant redwoods — otherwise carry straight on through the bright lime-green larches towards the Temperate House.

12L Fork left towards the Temperate House where the crab apple trees are in flower. Either detour through the house, or continue round the outside, but head left towards King William's Temple and the mass of blossom along Cherry Walk, planted with beautiful blue scillas under the trees.

13L At the crossroads after the temple, turn right (signpost Victoria Gate) and pass the Temple of Bellona, with a massed carpet of crocuses all around in February and March.

14L Carry on down this path and, as you reach the T-junction by the Kew Road wall, look right at the camellias on the way to the Marianne North Gallery with its superb collection of flower paintings. Go and admire them, or turn left back to the Victoria Plaza.

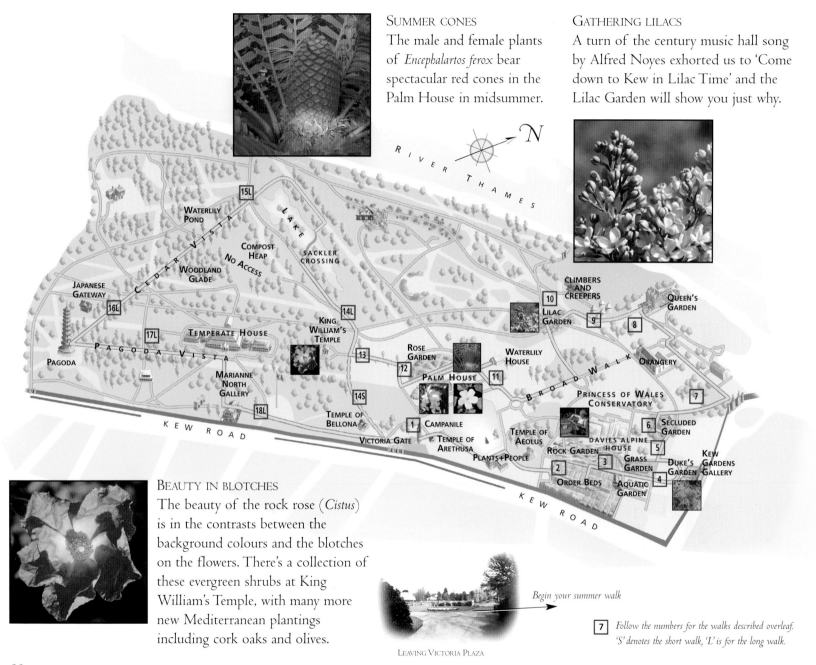

SUMMER CONES
The male and female plants of *Encephalartos ferox* bear spectacular red cones in the Palm House in midsummer.

GATHERING LILACS
A turn of the century music hall song by Alfred Noyes exhorted us to 'Come down to Kew in Lilac Time' and the Lilac Garden will show you just why.

RIVER THAMES

N

BEAUTY IN BLOTCHES
The beauty of the rock rose (*Cistus*) is in the contrasts between the background colours and the blotches on the flowers. There's a collection of these evergreen shrubs at King William's Temple, with many more new Mediterranean plantings including cork oaks and olives.

LEAVING VICTORIA PLAZA

Begin your summer walk

7 *Follow the numbers for the walks described overleaf.
'S' denotes the short walk, 'L' is for the long walk.*

Map labels:
WATERLILY POND, LAKE, COMPOST HEAP, NO ACCESS, SACKLER CROSSING, 15L, CEDAR VISTA, WOODLAND GLADE, JAPANESE GATEWAY, 16L, 17L, TEMPERATE HOUSE, PAGODA VISTA, PAGODA, MARIANNE NORTH GALLERY, 18L, KEW ROAD, KING WILLIAM'S TEMPLE, 14L, 13, 14S, TEMPLE OF BELLONA, VICTORIA GATE, 1, CAMPANILE, TEMPLE OF ARETHUSA, PLANTS+PEOPLE, ROSE GARDEN, 12, PALM HOUSE, 11, WATERLILY HOUSE, TEMPLE OF AEOLUS, ROCK GARDEN, DAVIES ALPINE HOUSE, 2, ORDER BEDS, 3, GRASS GARDEN, AQUATIC GARDEN, 4, DUKE'S GARDEN, CLIMBERS AND CREEPERS, 10, LILAC GARDEN, 9, 8, QUEEN'S GARDEN, BROAD WALK, ORANGERY, PRINCESS OF WALES CONSERVATORY, 6, SECLUDED GARDEN, 5, 7, KEW GARDENS GALLERY, KEW ROAD

Summer at Kew

Seasons and flowering times can vary by up to three weeks, so the periods below are only approximate.
For up-to-date details of flowering, please call 020 8332 5655 or visit the website at www.kew.org

From June to early September, the Gardens are in full bloom. Early on, the Lilac Garden is a picture, while the Secluded Garden, Duke's Garden and Queen's Garden make stunning, longer-lasting shows.

Magnificent Indian horse chestnuts display their huge candles of bloom, and the Rose Garden and Grass Garden are 'musts' for many visitors. Summer scents fill the air from the special selection of plants around King William's Temple. You'll find some clever ideas for dry gardening here and in the Duke's Garden.

The map shows you where the summer displays are, and if you would like to try a favourite summer walk, turn to the next page.

DRUNK WITH PLEASURE
Hibiscus - there are some in the Palm House - are wonderfully showy, though the flowers are usually short-lived. If you like drinking herbal teas, look at the ingredients - you'll find hibiscus prominent.

SUMMER SCENTS
The sweet perfume of frangipani (*Plumeria rubra*) and white spider lilies can fill the air in the Palm House. For more scents, follow your nose to the Rose Garden and the *Philadelphus* Collection by the Pagoda.

GIANTS
Giant waterlilies in the Princess of Wales Conservatory and the Waterlily House are as amazing now as they were to the early Victorians when the first specimens arrived here.

LAVENDER PINK, LAVENDER BLUE
The Lavender Trail at the Duke's Garden shows the wonderful variety of this popular plant, and includes French lavender with its attractive ear tufts. Shown here is *Lavandula stoechas* 'Kew Red'.

Summer Walks

You can take a long or a short version of this walk. At a gentle pace, stopping to look at plants, but not lingering, the short walk takes about an hour to an hour and a half; the long walk about an hour longer, starting and ending at the Victoria Plaza. You can follow the walk, which is on both paths and grass, from the description below and/or from the summer map on the previous page.

Rudbeckia

Dahlia

Japanese Minka

1 Go through the Victoria Plaza, pass the Campanile, turn right at the Temple of Arethusa and make your way round the Pond to the Plants+People Exhibition, where you will see just how important plants are to our lives. Turn sharp right at the end of the building and with the Temple of Aeolus on its hillock to the left, follow the path to the peony beds.

2 Past the peonies are the Order Beds, where related plants are grouped together in a colourful parade. Kew students have vegetable plots here and are awarded marks for them throughout the year.

3 Leave the Order Beds by the centre left exit and you are faced with the colourful Rock Garden. Head right, through the centre of the Rock Garden to the award-winning Davies Alpine House.

4 The Grass Garden has over 550 species on view, from fine lawn through cereals to bamboos. Next to it is the Duke's Garden with the sweetly-scented Lavender Trail and the Gravel Garden, full of ideas for plants that need very little water.

5 Turn right out of the Duke's Garden and on the left there's a path to the Princess of Wales Conservatory. If you have time, enjoy the seasonal display where you enter and a magnificent giant waterlily which should be in full flower in the centre of the house.

6 Continue along the path to the huge stone pine tree. Opposite is a sign leading into the Secluded Garden, with plants to stimulate all the senses, a conservatory and a bamboo tunnel.

7 Go through the Secluded Garden and turn right to the Main Gate, where you then turn left along a path lined with magnificent Indian horse chestnut trees, with candles of flowers proud on their branches (June).

8 Pass the rear of the Orangery, turn right and pass the Kew Palace Welcome Centre.

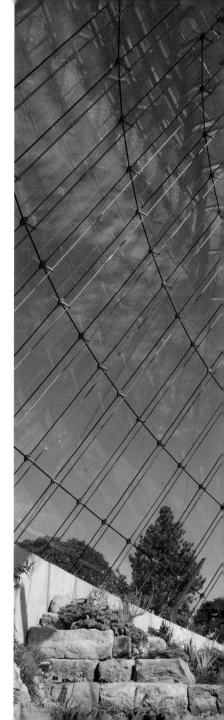

Kew Palace is well worth a visit. The elegant Queen's Garden, with its 17th century inspired parterre, a nosegay garden and a splendid laburnum arch, is behind the Palace.

9 Leaving Kew Palace, cut diagonally to the right across the grass to White Peaks, where you might stop for something to eat and drink, or visit Climbers and Creepers, a perfect break for children.

10 Leaving White Peaks, head straight on to the Lilac Garden (especially in early summer) and continue on to join the Broad Walk.

11 Turn right along the Broad Walk and right again at the roundabout for the Palm House. Turn right into the Waterlily House to see a giant waterlily, loofahs, papyrus and the sacred lotus.

12 Inside and either side outside the Palm House are joys. There's summer bedding on the Pond side and a vast display of roses and lavender on the other. Enjoy any or all of them, then take the path away from the roses in the direction of the Temperate House

13 Soon, there's a crossroads in front of King William's Temple, with its specially planted collection of highly-scented Mediterranean style shrubs, herbs and other plants.

Short walk or long walk?

Here is where you make your decision. For the short walk - you've almost completed it by now, with only about ten minutes to go - follow the sign to Victoria Gate. For the long walk, (around an hour more back to the Victoria Plaza), follow the signpost to the Lake.

Short walk

14S Your way back goes past the Temple of Bellona straight down to the Victoria Plaza where you started this summer day here in Kew Gardens. Stop here for a well-earned drink and to explore the shops.

Long walk

14L From the crossroads turn right until you reach where five paths meet, then continue virtually straight ahead to the Lake, taking the path by the lakeside. Just over halfway along, take the path on the right to the elegant Sackler Crossing. Cross the Lake and turn left onto Syon Vista. From here it is only a short detour to explore the Bamboo Garden and Japanese Minka if you have time. Otherwise, head up Syon Vista towards the river for some great views, and then walk around the top of the Lake.

short walk ← → long walk

15L From the head of the Lake, go over a grass hummock, across the path and down Cedar Vista towards the Pagoda. On the right, the Waterlily Pond is very attractive in summer and just past it on the left, the seclusion of the beautifully planted Woodland Glade makes a welcome diversion.

16L Carry on down Cedar Vista over several paths to the tranquillity of the Japanese Gateway and Landscape on the right and then, just ahead, the Pagoda reaches up to its full ornate height. There's a fragrant planting of *Philadelphus* by the Pagoda, filling the air with orange blossom scent.

17L Turn left and walk the length of the Pagoda Vista - don't forget to look behind you to appreciate the view. At the Temperate House turn right onto a path towards the Marianne North Gallery with its collection of flower paintings.

18L At the T-junction turn left onto the path that takes you back to Victoria Plaza.

THE MAIDENHAIR TREE
Why is *Ginkgo biloba* called the 'maidenhair tree'? Its attractive foliage looks very like that of the maidenhair fern. (See p 58.)

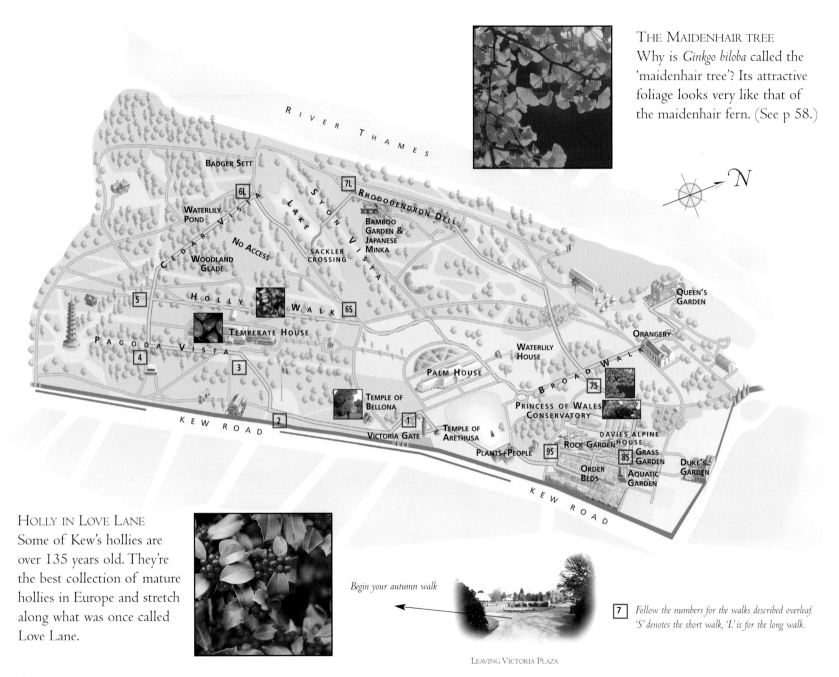

RIVER THAMES

BADGER SETT

6L

7L
RHODODENDRON DELL

WATERLILY POND

CEDAR VISTA

LAKE

SYON VISTA

NO ACCESS

BAMBOO GARDEN & JAPANESE MINKA

WOODLAND GLADE

SACKLER CROSSING

HOLLY WALK

5

6S

QUEEN'S GARDEN

PAGODA VISTA

TEMPERATE HOUSE

ORANGERY

4

3

WATERLILY HOUSE

BROAD WALK

7S

TEMPLE OF BELLONA

PALM HOUSE

PRINCESS OF WALES CONSERVATORY

1

2

VICTORIA GATE

TEMPLE OF ARETHUSA

DAVIES ALPINE HOUSE

ROCK GARDEN

PLANTS+PEOPLE

9S

8S

GRASS GARDEN

DUKE'S GARDEN

ORDER BEDS

AQUATIC GARDEN

KEW ROAD

KEW ROAD

HOLLY IN LOVE LANE
Some of Kew's hollies are over 135 years old. They're the best collection of mature hollies in Europe and stretch along what was once called Love Lane.

Begin your autumn walk

7 *Follow the numbers for the walks described overleaf. 'S' denotes the short walk, 'L' is for the long walk.*

LEAVING VICTORIA PLAZA

Autumn at Kew

Seasons and flowering times can vary by up to three weeks, so the periods below are only approximate.
For up-to-date details of flowering, please call 020 8332 5655 or visit the website at www.kew.org

September and October and, so variable are our seasons nowadays, probably well into November, bring in the poetic 'season of mists and mellow fruitfulness'. Autumn colour, as deciduous trees shut down their leaf activity, and berries ripen, is at its most striking.

There's still more colour with autumn crocus, hardy cyclamens and belladonna lilies, while the Grass Garden never ceases to fascinate with its huge variety of plants, ranging from the highly decorative ornamental grasses to the most useful cereal crops. If ever a plant family supported mankind, it's the grasses. Just three cereals - maize, wheat and rice - provide nearly two-thirds of the calories and half the protein consumed by the world's population.

The Woodland Glade and the Duke's Garden are also thoroughly rewarding to visit at this time of year.

SMOKE, BUT NO FIRE
The smoke bush (*Cotinus obovatus*) is famous for its hazy covering of feathery flowers that gives it its name. Its splendid autumn colours are almost luminous.

AN ANCIENT INHABITANT
This splendidly gnarled Japanese pagoda tree (*Styphnolobium japonicum*) is one of Kew's oldest trees, having been brought here in 1762. Find out more about Kew's heritage trees and their places in the Gardens' history on pp 58 & 59.

HOT STUFF
Chillies (*Capsicum*), shown here, are fruiting in the Temperate House, along with citrus fruits and the curious tree tomatoes.

Autumn Walks

There are long and short versions of this walk, which starts and ends at the Victoria Plaza. The short walk can take between 90 minutes and two and a half hours, depending on how much time you spend in the glasshouses and on detours. The long walk takes about an hour longer. You can follow the walks, which are on both paths and grass, from the description below and/or from the autumn map on the previous page.

BLAZING MAPLES
A touch of America's eastern seaboard comes to Kew with a blaze of 'fall colour'.

THE PAGODA
The Pagoda was built for Princess Augusta, who founded the botanic garden at Kew in 1759.

1 Exit the Victoria Plaza near the ticket booths and turn left onto a path parallel to Kew Road, where the smoke bush is changing colour by the Temple of Bellona.

2 Continue along the path, turn right and walk through a blaze of autumn colour among the maples towards the Temperate House.

3 Before you reach it, turn left on to the grass of Pagoda Vista, and stroll through more glorious autumn colour.

4 Turn right at the first path (away from the Pavilion Restaurant), and go past the end of the Temperate House to a T-junction.

SPECTACULAR SPINDLE TREES
Spindle trees (*Euonymus*) are grown for the spectacular colours of their autumn fruits, in which the seeds are just as brightly coloured. The evergreen varieties are good seaside plants.

16

5 Short walk or long walk?

Here is where you make your decision. For the short walk (one or two hours back to the Victoria Plaza, depending on the stops you make), turn right at the signpost to the Palm House and Evolution House. For the long walk (about another hour longer), continue straight up, over the grass, to where the broad spread of Cedar Vista forks off to the right.

long walk

short walk

Short walk

6S It's burgeoning berries all the way as you stroll parallel to the Temperate House along Holly Walk. At a five-way junction, fork right and then turn right on the path through the Rose Garden and then between the Palm and Waterlily Houses. Carry on to the roundabout and turn left up the Broad Walk towards the Orangery. Here, you can make a detour to the Orangery for light refreshments and, if you have the time (about an extra 20-30 minutes), you could see the beautiful autumn-flowering cyclamen under the pleached hornbeams in the Queen's Garden behind Kew Palace. ('Pleaching' forms a high hedge on tall stems.)

7S Otherwise, halfway down the Broad Walk turn right, off towards the Princess of Wales Conservatory and the Japanese pagoda tree is on the right. Turn right at the historic urn and head to the Princess of Wales

Conservatory with its seasonal display in the north end. Beyond the Conservatory you can visit the Davies Alpine House on one side and the Grass Garden on the other.

8S Head through the Davies Alpine House and take the path through the Rock Garden. At the end of the main path turn left to the Order Beds, which show just how attractive serried ranks of seed heads can look.

9S Now head right towards Museum No I and the Plants+People Exhibition, which is really worth while seeing, so try to make the time. To end your autumn walk, continue round the Pond opposite the Palm House and turn left at the Temple of Arethusa to the Victoria Plaza.

Long walk

6L From 5 above, you have continued over the grass to fork right onto Cedar Vista, walking through glorious autumn colour, past the Waterlily Pond. Don't forget to look back at the view. At the end of Cedar Vista you meet a path, where you turn right. When you reach the Sackler Crossing, stroll over to enjoy the autumnal reflections in the Lake. On the other side, cross Syon Vista, through the trees towards the Bamboo Garden and Rhododendron Dell.

7L In the centre of the Bamboo Garden is the intriguing Japanese Minka. Take the central path of the Rhododendron Dell and then carry straight on to another path, where you turn right to reach the Broad Walk. There, a left turn to the Orangery would bring you some light refreshments. And if you have a little extra time (about 20-30 minutes), you could see the beautiful autumn-flowering cyclamen in the Queen's Garden behind Kew Palace. Otherwise, cross straight over to join up with the short walk route at 7S.

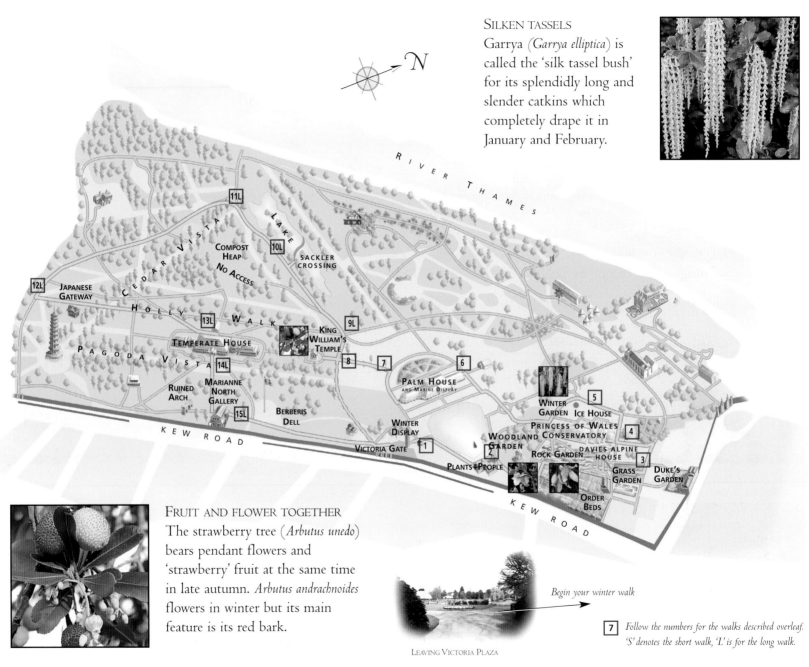

SILKEN TASSELS
Garrya (*Garrya elliptica*) is called the 'silk tassel bush' for its splendidly long and slender catkins which completely drape it in January and February.

N

RIVER THAMES

11L
10L
12L
JAPANESE GATEWAY
COMPOST HEAP
NO ACCESS
LAKE
SACKLER CROSSING
CEDAR VISTA
HOLLY WALK
13L
TEMPERATE HOUSE
PAGODA VISTA
14L
KING WILLIAM'S TEMPLE
9L
8
7
6
RUINED ARCH
MARIANNE NORTH GALLERY
15L
BERBERIS DELL
PALM HOUSE
AND MARINE DISPLAY
WINTER GARDEN
ICE HOUSE
5
PRINCESS OF WALES CONSERVATORY
4
WINTER DISPLAY
WOODLAND GARDEN
DAVIES ALPINE HOUSE
3
VICTORIA GATE
1
ROCK GARDEN
GRASS GARDEN
DUKE'S GARDEN
KEW ROAD
PLANTS+PEOPLE
ORDER BEDS
KEW ROAD

FRUIT AND FLOWER TOGETHER
The strawberry tree (*Arbutus unedo*) bears pendant flowers and 'strawberry' fruit at the same time in late autumn. *Arbutus andrachnoides* flowers in winter but its main feature is its red bark.

LEAVING VICTORIA PLAZA

Begin your winter walk

7 *Follow the numbers for the walks described overleaf.* 'S' *denotes the short walk,* 'L' *is for the long walk.*

Winter at Kew

Seasons and flowering times can vary by up to three weeks, so the periods below are only approximate.
For up-to-date details of flowering, please call 020 8332 5655 or visit the website at www.kew.org

November right round to February - and there's more to see in the Gardens than you might think. At the end of the year, there is winter bark to look out for; the pretty winter-flowering cherry and the simultaneous flowers and strawberry-like fruit of the strawberry tree. Try Holly Walk to ease you into the festive spirit.

The New Year brings winter-flowering shrubs, such as wintersweet and witch hazel, with heady scents and a mass of flowers, all round the Ice House; hellebores in the Order Beds; and a great show of snowdrops in the Rock Garden.

It's a comforting thought, visiting Kew in the depths of winter, that you can always warm up in one of the glasshouses and imagine you have escaped to the lush tropics.

HIDDEN BEAUTY
Low down near the ground, the nodding flowers of the hellebores are hiding their beauty. The Christmas rose (*Helleborus niger*) appears around midwinter.

A LITTLE LIVERISH?
Clematis cirrhosa is one of the few evergreen clematis. Its late winter flowers are speckled inside and look like a diseased liver - hence *cirrhosa*.

COMPOST - HEAPS OF INTEREST
Peat - once every gardener's first choice for potting and seed composts, mulching and soil improvement - is a rapidly dwindling natural resource and Kew largely suspended its use in 1989. Today, Kew mixes the 10,000 cubic metres of waste plant material it generates every year with horse manure from some rather distinguished stables: the Royal Horse Artillery, the Metropolitan Police and the Wardens of Windsor Great Park. After watering and turning through a 10-12 week cycle, it produces 2,000 cubic metres of

compost for use throughout the Gardens. There is a viewing platform in the Pinetum, close to the Lake, together with examples of how home composting can improve garden soil and reduce pressure on landfill sites - a worthy effort.

Winter Walks

You can take a long or a short version of this walk. At a gentle pace, the short walk takes about an hour, the long walk about two hours, starting and ending at the Victoria Plaza. You can follow the walk, which is on both paths and grass, from the description below and/or from the winter map on the previous page.

BEAUTIFUL BARK
Birch trees are known for their beautiful bark, which varies enormously in colour. This is silver birch (*Betula pendula*).

HARBINGERS OF SPRING
It's a sure sign that spring is on its way when the tough pointed buds of snowdrops (*Galanthus*) push their way through hard ground and open their demure little flowers.

1 On leaving the Visitor Plaza you can enjoy several trees with beautiful winter bark. Go past the Temple of Arethusa and turn right at the Palm House Pond along a glorious row of sweetly-scented winter-flowering viburnum to the Plants+People Exhibition, which is well worth a visit.

2 Turn right at the end of the building, and pass the Temple of Aeolus where you will find early daffodils on the mound if spring has come early, flowering hellebores, daphne and snowdrops in the Woodland Garden. Follow the path round past the Order Beds and, after the wall, turn right into the Rock Garden, taking its centre path.

3 Take this path to the award-winning Davies Alpine House and on to the Grass Garden and its sculptural seed heads. Go through the grasses to the path by the Duke's Garden where you'll find wintersweet. Turn left and go up the path to the stone pine and turn left again to the Princess of Wales Conservatory where there is an interesting seasonal display.

4 On turning right at the Conservatory there's a paper-bark maple on the corner, with stunning bark - dark brown peeling off to reveal layers of vibrant orange beneath (shown left).

5 Go up to the T-junction and turn left past the Japanese pagoda tree and keep left for the Ice House and Winter Garden with its fragrant flowers. Continuing round, turn right and head for the roundabout and then on to the Palm House. The Pond outside usually has a number of ornamental wildfowl preening themselves.

6 Go through the lush warmth of the Palm House, taking in the Marine Display in the basement for a fascinating glimpse of four marine environments; and out past Kew's oldest pot plant through the far (south) door and, on leaving, turn right.

7 Take the first path left towards King William's Temple, around which there are strawberry trees on the Lake side and a collection of witch hazels (in bloom from December to February) on the other, as well as many other intriguing plants.

8 *Short walk or long walk?* Here is where you make your decision. For the short walk, simply follow the signposts to Victoria Gate, which is very close.

Long walk

9L For the long walk (about an hour more from here), follow the signpost to the Lake.

10L Take the right fork to the Lake, on which you may well see some interesting waterfowl. Halfway down you will see the elegant Sackler Crossing, worth a detour for an excellent view of the Lake. Just beyond, on the left, is the path to the Compost Heap viewing platform.

11L At the head of the Lake, past Cedar Vista, take the path that forks left and follow the signs to Lion Gate and the Pagoda, which take you through the majestic giant redwoods.

12L Take the second left to pass the Japanese Gateway on your right. Keep straight ahead along Holly Walk until you fork right and go in through the middle doors of the Temperate House.

short walk   ←   →   long walk

13L Inside the Temperate House, you'll find the spectacular bird of paradise flowers in full bloom, along with *Banksia* and sub-tropical rhododendrons.

14L Leave by the middle doors on the Kew Road side, walk straight ahead towards the Marianne North Gallery, which is well worth your time to see the extraordinary collection of her flower paintings.

15L Turn left (signposted Victoria Gate) past the Berberis Dell with some eye-catching winter berries on show. A few yards further on is the Victoria Plaza where you started this winter walk.

The Palm House

DECIMUS BURTON'S IMPRESSIVE BUILDING, HOUSING ONE OF THE WORLD'S FOREMOST COLLECTIONS OF TROPICAL RAINFOREST PALMS AND THE MARINE DISPLAY

Palms are second only to grasses in their importance to people. About 70% of all palm species are found in tropical rainforests, one of the most threatened habitats on Earth. The Palm House creates conditions similar to tropical rainforest; around a quarter of the palms planted here are threatened in the wild, as are more than half of the cycads, the 'living fossils' of the tropics.

The Palm House also contains many plants of great economic significance, grown for their yields of fruits, timber, spices, fibres, perfumes and medicines. Plants are grouped together in geographical areas, except in the centre, where the tallest specimens need the extra height of the dome.

The Marine Display in the basement recreates four important marine habitats, complete with fish, corals and other sea creatures, and shows the importance of marine plants.

THE WORLD'S SMELLIEST FRUIT

The legendary durian has a sublime taste, but a foul smell - the experience has been described as 'Like eating custard in a sewer.'

THE WORLD'S LARGEST SEED

Coco-de-mer (*Lodoicea maldivica*) grows wild on only two islands in the Seychelles. Its enormous fruit takes seven years to develop and results in an unusual two-lobed seed, which can weigh up to 30 kg – the same as the average TV set! This is packed with food reserves to sustain the germinating plant.

AFRICA
South Wing

Of the very few African palm species, one - the African oil palm - is the most important oil-producing plantation palm in the tropics as well as one of the most controversial. Madagascar and other African islands have more palms and Kew is proud of its rare triangle palm and the distinctive *Beccariophoenix madagascariensis*.

Other valuable plants which originated in Africa include coffee bushes which often produce berries in the Palm House and the Madagascar periwinkle from which anti-leukaemia drugs were developed.

South Africa is rich in cycads (shown right) and one of Kew's specimens is our oldest pot-plant, having been brought here in 1775.

THE AMERICAS
Centre Transept

The Victorians concentrated on palms from the Americas, the world's richest rainforest habitat. Caribbean palms and rare Mexican cycads have their own areas and visitors often recognise their own houseplants in some of the *Chamaedorea* palms.

Many vitally important economic specimens are planted here, including cocoa, rubber, bananas, papaya, soursop, cherimoya and mammee-apple. The Mexican yam was used to develop the contraceptive pill and the tonic drink sarsaparilla is made from the roots of *Smilax utilis*.

The parrot flowers (*Heliconia*) (shown left) and the hybrid rose of Venezuela often display splashes of scarlet and in summer, the sweet scents of frangipani and white spider lilies can fill the air.

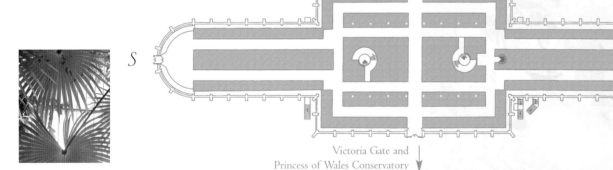

W

↑

Temperate House
and Lake

S

N

Victoria Gate and
Princess of Wales Conservatory

↓

E

THE TALLEST PALMS
Centre Transept

What dictates a planting position under the great central dome of the Palm House is not geography, but sheer height, which cannot be reduced without killing the growing point, or 'heart' of the plant. The many-stemmed peach palm is often felled for its edible heart - 'coeur de palmier'.

Also here are the babassu from central South America; the fast-growing queen palm, so popular with landscapers; the American oil palm and the pantropical coconut palm which actually fruits at Kew.

ASIA, AUSTRALASIA, THE PACIFIC
North Wing

This area of the world is home to the greatest diversity in the palm family and the rich collection here includes shade-loving 'understorey' specimens, dwarf palms and climbing rattans, from which most 'cane' furniture is made.

Many familiar Asian fruit trees are planted here, such as mango and starfruit, along with the more unusual breadfruit, jackfruit, and Indonesia's legendary durian. Sugar cane and common spices, such as ginger, pepper and cardamom, are seen here as growing plants.

Other Palm House notables include the spectacular jade vine with its enormous iridescent masses of jade green flowers and the wide selection of potted palms with their characteristic 'feather' or 'fan' leaves.

The Palm House

PLANTING AND CONSERVATION

The Palm House represents one habitat - tropical rainforest. The plantings simulate its multi-layered nature with canopy palms and other trees, climbers and epiphytes and then down to the shorter understorey plants and dwarf palms. The cycads here are solely from rainforests.

Many of the plants in this collection are endangered in the wild; some are even extinct. Their natural habitats are being destroyed by clearing for agriculture, mining, or logging. Scientists at Kew are deeply involved in surveys and studies of palms and other rainforest species to assess the diversity of habitats and to develop methods for the sustainable growth and harvesting of these plants, as well as their conservation.

All life depends on plants. If left unchecked, the current rate of destruction of the world's rainforests will see the total loss of this environment by the middle of this century, together with its peoples, its vast untapped potential for crops and medicines, and its vital moderating effect on the Earth's climate.

The Palm House, then, is more than just a collection to gaze at and wonder. It is an invaluable world resource.

RAINFOREST PLANTS IN THE WORLD ECONOMY

The Palm House collection contains many plants of significant economic importance, and one of Kew's roles is research into many of the factors that make for sustainable cropping.

Here are some examples to look for:-

RUBBER TREE
(*Hevea brasiliensis*)

Natural rubber has unique properties, making it important for special uses, from robust space shuttle tyres to delicate contraceptives. In 1876, Kew received 70,000 seeds collected from the tree's native Amazonian Brazil. Only 2,800 germinated, but from them, the seedlings sent to Sri Lanka and Malaysia flourished and started their rubber industries.

Rubber on tap
The white latex that flows in the inner bark of the rubber tree is tapped by a series of cuts.

AFRICAN OIL PALM
(*Elaeis guineensis*)

Native to tropical Africa, this, the most productive oil-producing tropical plant, is grown in plantations throughout the world's humid tropics. Palm oil pressed from the orange flesh makes soap, candles and some edible products. Oil from the stone's hard white kernel is also used for edible oils and fats, soaps and detergents. In Africa, the sap is tapped for palm wine, and the leaves and trunk help make houses.

COCOA (*Theobroma cacao*)

The Aztecs called it 'Food of the Gods', hence its scientific name deriving from 'theos' (gods) and 'bromos' (food). Originally a bitter beverage reserved for Aztec high society, the Spanish added sugar and vanilla to create its modern taste.

GIANT BAMBOO
(*Gigantochloa verticillata*)

Bamboos are in the grass family and are known as 'friend of the people' in China and 'wood of the poor' in India. Giant bamboo is highly versatile, used in house construction and paper-making, while the young shoots are a table delicacy. However, land clearance for agriculture has resulted in the total destruction of many natural habitats of bamboo.

PEPPER (*Piper nigrum*)

The fruits of this woody climber, native to India, are the familiar peppercorns, yielding the world's most important spice. Black pepper comes from grinding the whole dried peppercorn; for white pepper, the outer husk is removed. Pepper was first used as a cooking spice in India; then came to Europe in the Middle Ages to flavour and cure meat. It was also used with other spices to mask the taste of bad food.

MADAGASCAR PERIWINKLE
(*Catharanthus roseus*)

Traditionally used in folk remedies for digestive complaints and diabetes, this pretty ornamental (shown far left) is recognised as important in the fight against cancer. Two of its six especially useful alkaloids can be used to treat leukaemia and Hodgkin's Disease. Commonly cultivated, it is now rare in the wild as deforestation destroys its habitat and that of other potentially useful relatives.

COFFEE (*Coffea*)

Coffee was first used as a paste of beans and oil, chewed for the caffeine effect, but by the 15th century people were roasting beans to produce a drink. Originally African, most of the world's coffee now comes from Latin America. Arabica beans (*Coffea arabica*) make quality coffee, while the robusta (*Coffea canephora*) is a better commercial crop generally used for instant coffee.

COCONUT
(*Cocos nucifera*)

The origin of the most familiar tropical palm is probably in the islands of the western Pacific and eastern Indian oceans, spreading by ocean currents and human planting. Every part of the tree is used; the trunk for timber; leaves provide thatch, baskets and brooms; while the nut provides food, drink and kernel oil from inside and coir fibre and garden mulch from the outside.

SUGAR CANE (*Saccharum officinarum*)

Another invaluable grass, sugar cane originated in New Guinea, where ancient people used the stem for chewing. Today, raw cane sugar is a major world commodity and by-products such as molasses and bagasse (used for fuel) are valuable, too. Fuel alcohol produced from sugar is used in 50% of all cars in Brazil - a cheap, renewable and less polluting tankful.

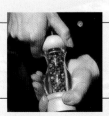

PEPPERCORN RENT
Peppercorns were once so expensive that rents were paid in them - hence peppercorn rent. Today, it means just a token payment.

The Palm House

The icon of Kew

Built 1844-48 by Richard Turner to Decimus Burton's designs, the Palm House is Kew's most recognisable building, having gained iconic status as the world's most important surviving Victorian glass and iron structure.

Borrowed technology

While the thinking was Burton's, the *doing* - the extraordinary engineering and construction work - was very much Richard Turner's. The technology was borrowed from shipbuilding and the design is essentially an upturned hull. The unprecedented use of light but strong wrought iron 'ship's beams' made the great open span possible.

Marrying form and function

The Palm House was built for the exotic palms being collected and introduced to Europe in early Victorian times. The elegant design with its unobstructed space for the spreading crowns of the tall palms was a perfect marriage of form and function.

Burton chose the site so that his building would be reflected in the Pond to the east. Two tree-lined vistas - Pagoda Vista and Syon Vista - landscaped by William Nesfield, radiate from this focal point of the Gardens.

Campanile

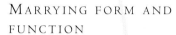

Originally, the Palm House boilers were in the basement, heating water pipes under iron gratings on which the plants stood in great teak tubs, or in clay pots on benches. The smoke from the boilers was led away through pipes in a tunnel to an elegant Italianate campanile smoke stack 150 m (490 ft) away. However, the boiler room was constantly flooded, and some parts of the Palm House were too cold.

By the early 1950s, over a hundred years of external weathering, together with high temperature and humidity inside, had taken their toll. Restoration was long overdue.

The World's Oldest Pot-Plant?

This cycad, *Encephalartos altensteinii*, was brought to Kew in 1775 and typifies the Palm House's original arrangement of plants displayed in pots. It needed special care when moved for the restoration in the 1980s.

The Restorations

The first restoration was in 1955-57, when glazing bars were cleaned and realigned and the house completely reglazed. New beds were built and the plants simply moved around to accommodate the works.

The boilers were converted to oil and moved to a site behind the Campanile, using the tunnel to take hot water to the heating system. A later conversion to dual-fuel (oil or gas) boilers has resulted in gas being the standard fuel.

PALMS ON THE MOVE

The second restoration, 1984-88, was much more comprehensive after major work was found to be vital, for both the building and the heating system.

For the first time in its history, the Palm House was completely emptied. Most palms were held in temporary glasshouses, but some were too large, so were cut down and made into specimens for the Herbarium and Museum.

GRADE I LISTED BUILDING

The Palm House is a world-famous Grade I listed building. It was dismantled entirely, restored and rebuilt, with parts replaced like for like. Burton and Turner were at the leading edge of their glazing technology, but standards today are vastly superior. Toughened safety glass is held by ten miles of glazing bars made of modern stainless steel, but to exactly the same section as the originals.

The floor layout was revised to allow planting in beds, rather than pots, which made room for wider paths and seats for visitors. The basement was enlarged to house staff facilities and the new Marine Display.

The second restoration took as long to complete as the original house took to build. Replanting was completed in mid-August 1989 and H.M. Queen Elizabeth the Queen Mother officially reopened the Palm House on 6 November 1990.

The Marine Display

All life originated in the sea. Today, millions of years later, life still depends on the most simple marine plants - the algae. Algae, which include all seaweeds, provide half of the world's oxygen supplies and absorb vast amounts of carbon dioxide. As Sir David Attenborough put it, "Without algae, there would be no life on earth, the seas would be sterile and the land would be uncolonised."

The Marine Display emphasises the importance of marine plants and, through displays in 19 tanks, recreates four major marine habitats: coral reefs, estuaries and salt marshes, mangrove swamps and rocky shorelines.

WHY SEAWEEDS ARE DIFFERENT COLOURS
Seaweeds are divided into groups by colour - green, brown and red. Green seaweeds live in shallow waters up to 10 m deep. Brown algae can survive in water up to 30 m deep, while red algae can survive in water to a depth of 100 m. All algae need light for photosynthesis, and each group contains different coloured pigments to absorb the available daylight at different depths.

EAT AND BE EATEN
Algae are a vital first link in the ocean's food chain. They are also largely unexploited by man, but that situation is changing. Japanese nori and Welsh laverbread are well-established foods. Seaweed has been widely used as land fertiliser and animal fodder in coastal communities, too. Today, algal products include a variety of gels, such as agar, alginates and carrageenan; used in ice cream, growing media for microbial culture, paints and very effective wound dressings.

Seaweed

LITTLE AND LARGE
Algae vary enormously in shape and size. The tiny, single-celled micro-algae (also called phytoplankton), include the world's smallest plant, *Chlorella*, which is only 0.003 mm in diameter. Among the huge macro-algae is giant kelp (*Macrocystis*) which, at up to 100 metres in length, is among the world's largest plants.

Giant kelp

GREEN SCUM AND RED TIDES
Micro-algae as single cells are usually invisible to the naked eye, but occur in vast numbers as green scum on ponds and red tides on beaches and in reservoirs. Under a microscope, they show a huge variety of often very beautiful shapes. Blue-green algae (cyanobacteria) are the world's oldest plants, appearing some 2,000 million years ago.

Tube Coral

10 metres

30 metres

100 metres

The Temperate House

THE LARGEST GLASSHOUSE AT KEW, FULL OF TENDER WOODY PLANTS FROM SUBTROPICAL AND WARM TEMPERATE REGIONS WORLDWIDE

Once the largest plant house in the world and now the world's largest surviving Victorian glass structure, the Temperate House is another of Decimus Burton's designs and yet another of Kew's 39 listed buildings. At 4,880 square metres, it is the largest public glasshouse at Kew, twice the size of the Palm House.

Tender woody plants from the world's temperate regions have always been a major part of the collection at Kew. In Victorian times, the intensity of collecting meant that the Orangery and many other houses quickly became vastly overcrowded so, in 1859, it was decided to build another major glasshouse to complement the Palm House.

Today, the planting is in geographical zones as intended in Burton's original design. The scheme now represents many more regions than there were originally.

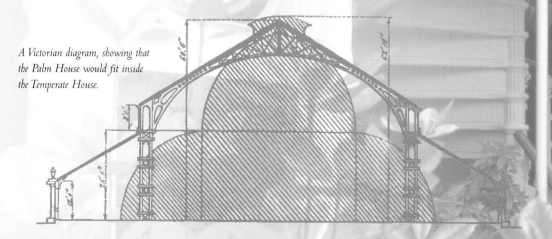

A Victorian diagram, showing that the Palm House would fit inside the Temperate House.

THE WORLD'S LARGEST INDOOR PLANT is the Chilean wine-palm (*Jubaea chilensis*) in the centre of the Temperate House, which is 16 m (52 ft) high - and still growing! It was grown from seed and there is a replacement nearby, ready for the time when this huge wine palm no longer fits into the roof space.

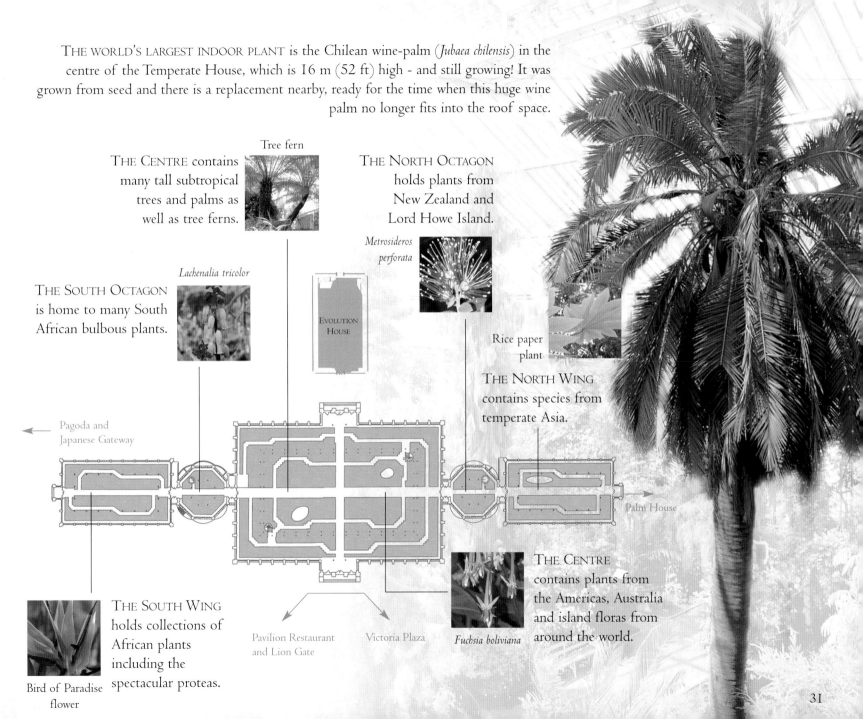

Tree fern

THE CENTRE contains many tall subtropical trees and palms as well as tree ferns.

THE NORTH OCTAGON holds plants from New Zealand and Lord Howe Island.

Metrosideros perforata

Lachenalia tricolor

THE SOUTH OCTAGON is home to many South African bulbous plants.

EVOLUTION HOUSE

Rice paper plant

THE NORTH WING contains species from temperate Asia.

Pagoda and Japanese Gateway

Palm House

THE SOUTH WING holds collections of African plants including the spectacular proteas.

Pavilion Restaurant and Lion Gate

Victoria Plaza

Fuchsia boliviana

THE CENTRE contains plants from the Americas, Australia and island floras from around the world.

Bird of Paradise flower

31

The Temperate House

TEMPERATE ZONES - TENDER WOODY PLANTS, MANY OF GROWING ECONOMIC IMPORTANCE

The plant collection in the Temperate House includes many spectacularly beautiful specimens that are deservedly admired, but it represents much more than that. Among the plants on display here are endangered island species being propagated for reintroduction to their native lands, such as *Hibiscus liliiflorus* from Rodrigues Island and *Trochetiopsis erthroxylon* from St. Helena.

There are also many plants of significant economic importance such as the date palm, tea, quinine and chillies.

The Temperate House holds an extensive collection of temperate American plants, including fuchsias, salvias and brugmansias. Also in the central section of the Temperate House is the Australian collection, with grass trees; the delightful 'kangaroo's paws' (so called for the shape of its flowers), and a fine array of banksias, named after Joseph Banks, who collected them and who is so intimately connected with Kew.

THE RAREST PLANT AT KEW

A cycad, *Encephalartos woodii*, was presented to Kew by the Natal National Park and is not only the rarest plant in the Gardens, but one of the last surviving specimens in the world. All cycads bear male (pollen) and female (seed) cones on different plants. The Kew tree is a lone male and it remains an extraordinary challenge to encourage it to cone.

FLOWERS AFTER 160 YEARS

This king protea (*Protea cynaroides*) from the Cape seems to have relished the conditions after the 1980 restoration, as it bloomed in 1986 after a gap of 160 years and can be seen growing in the south end of the House.

'G & T' TO LIFE-SAVER
From the bark of the cinchona
tree comes quinine, famously
used as a flavouring in Indian
Tonic Water and taken world
wide as a life-saving drug to
counter malaria.

33

The Temperate House

A PROUD HISTORY

In the mid-nineteenth century, the need for a large temperate greenhouse had become overwhelming, as the collection of tender woody plants had become so large. In 1859, the Government allocated £10,000 to build the Temperate House and directed Decimus Burton to prepare designs for this 'long-desiderated' conservatory. The Treasury called a halt when the account from Messrs Cubitt & Co for the construction of the main block and octagons came to £29,000. Overspending on Government projects is not, it seems, a modern phenomenon.

CREATING THE LAKE

The chosen site in the then-new Arboretum was raised two metres by creating a huge terrace of sand and gravel excavated from, and creating, the Lake site. The plants needed good ventilation, so the house was designed in straight lines. The glazing bars were of wood, not iron, for easy repair and to aid heating, and the aesthetics of the day dictated a decorated cornice at the eaves.

STOP-START BUILDING

Work began in 1859. The octagons were completed in 1861, the centre section in 1862 and foundations for the wings were part laid when work was stopped by the Treasury in 1863. Work was not resumed until more than 30 years later, in August 1895. The south wing was finished in 1897, then the contractor became bankrupt, so the north wing was completed by another in 1899.

CONTINUOUS PLANTING

As soon as the octagons were ready, the entire contents of the 'New Zealand House' were transferred to them. This had been the 'Great Stove', Kew's first hothouse, built a hundred years earlier, which was demolished as soon as the plants were removed. Its last trace is found near the Princess of Wales Conservatory, in the shape of an iron frame supporting a fine display of wisteria which once draped the building.

Australian plants, tubs of 'unhappy trees' from the Orangery, and some palms from the now-crowded Palm House were all soon established in their new beds. The central interior had 20 oblong beds, with lines of araucarias, palms, tree-ferns and other tall plants in the middle, side beds of rhododendrons, acacias, camellias and magnolias, with smaller plants in pots and boxes on benches. The admiring public were first admitted in May 1863, when the Temperate House was barely two-thirds finished.

THE LATER YEARS

In 1977, a little over a hundred years later, a full restoration was started. The Temperate House is a Grade I listed building, so scrupulous care was taken to keep the integrity of the original design. Just as Burton had used the best materials available to him, so the renovation used the best available - an aluminium and neoprene glazing system - but kept faithful to his design and the rhythm of the original sashes.

Modern heating technology keeps the building frost free - about 6-7°C - over the winter. It uses heat exchangers situated in the basements of the octagons, with intense heat piped from the main boilerhouse a quarter of a mile away.

A NEW LEASE OF LIFE

Almost 150 years after work commenced, the Temperate House stands proud as another of Kew's iconic buildings, housing a fascinating and important collection. It also provides a dramatic backdrop to Kew's immensely popular annual Summer Swing music festival and the lively Ice Rink in winter.

The Princess of Wales

KEW'S MOST COMPLEX GLASSHOUSE, WITH TEN DIFFERENT COMPUTER CONTROLLED ENVIRONMENTS IN ONE BUILDING

Opened by Diana, Princess of Wales on 28 July 1987, this most complex of Kew's public glasshouses commemorates Princess Augusta who married Frederick, Prince of Wales, in 1736 and who founded the Gardens.

Ten different environments cover the whole range of conditions in the tropics, ranging from scorching arid desert to moist tropical rainforest - all under one roof. Water features strongly in the humid zones, with pools of impressively large fish and the famous giant Amazonian waterlily.

Plants of great economic importance grow here too, such as pepper, bananas and pineapples, as well as orchids and carnivorous plants.

OPEN WIDE!
Insects are attracted to the wide open mouth of the pitcher plant by glistening droplets of nectar. But as soon as they climb inside to reach it, they lose their footing on the slippery walls and plop into a pool of digestive juices at the bottom of the 'pitcher'. But some insects have learned to retaliate, because their larvae are not only immune to, and swim around in, that strong digestive fluid, but actually live on scraps of food the plant does not use.

A HUGE ATTRACTION
The giant Amazonian waterlily (*Victoria amazonica*) was a huge attraction in Victorian times. A close relative (*Victoria* 'Longwood hybrid') is just as popular today, with its leaves measuring a full 2 m (6 ft 6 ins) across and flowers that change colour from white to deep pink over 24-36 hours.

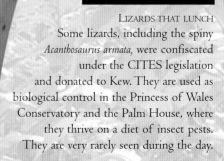

LIZARDS THAT LUNCH
Some lizards, including the spiny *Acanthosaurus armata*, were confiscated under the CITES legislation and donated to Kew. They are used as biological control in the Princess of Wales Conservatory and the Palm House, where they thrive on a diet of insect pests. They are very rarely seen during the day.

Conservatory

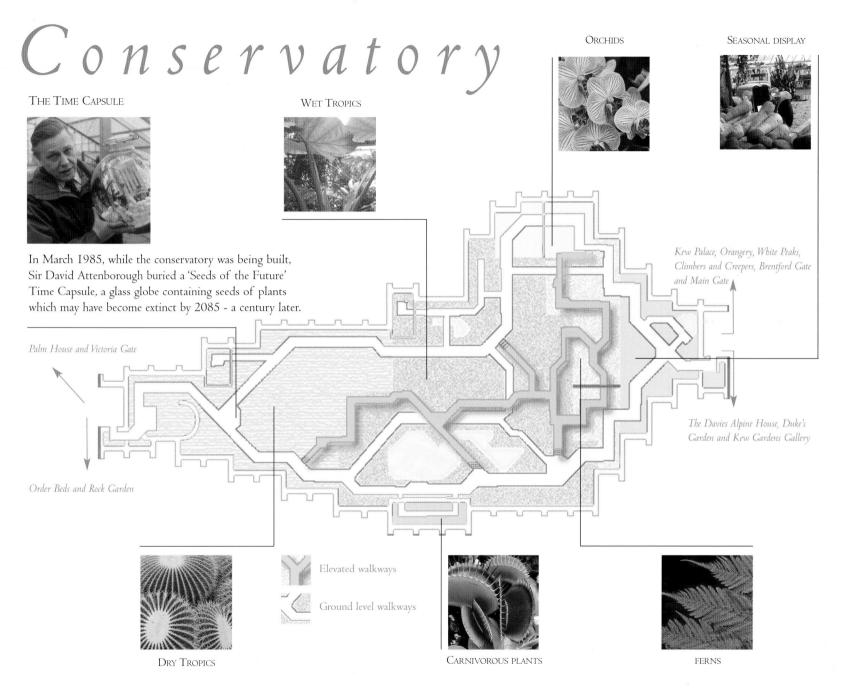

THE TIME CAPSULE

In March 1985, while the conservatory was being built, Sir David Attenborough buried a 'Seeds of the Future' Time Capsule, a glass globe containing seeds of plants which may have become extinct by 2085 - a century later.

WET TROPICS

ORCHIDS

SEASONAL DISPLAY

Kew Palace, Orangery, White Peaks, Climbers and Creepers, Brentford Gate and Main Gate

Palm House and Victoria Gate

Order Beds and Rock Garden

The Davies Alpine House, Duke's Garden and Kew Gardens Gallery

Elevated walkways

Ground level walkways

DRY TROPICS

CARNIVOROUS PLANTS

FERNS

The Princess of Wales

FROM WET TROPICS TO ARID LANDS, AND A VARIETY IN BETWEEN, EXPLORE A WHOLE NEW WORLD OF EXOTIC AND ECONOMIC HERBACEOUS PLANTS

Two main climate zones, the Dry Tropics and Wet Tropics, occupy most of this conservatory. There are eight more different micro-climates in the conservatory, each created for the special needs of a particular plant group. All plants are shown as naturalistically as possible, with ferns clinging to dripping rock faces, and climbers on columns. Paths on different levels bring visitors close to the plants so they can appreciate the subtle details of the vegetation.

THE DRY TROPICS ZONE

This zone represents arid regions from around the world. Here, there are plants that have adopted many different ways of dealing with the lack of water in their environment, conserving it with waxy skins or fleecy jackets, or storing it in succulent stems. Some of the cacti and agaves are displayed against a painted diorama, forming an atmospheric desert backdrop.

Although coming from opposite sides of the world, the agaves and aloes nearby show many similarities in adaptations for survival in these conditions.

There are also collections of the unique native plants from the Canary Islands and Madagascar, many of which are threatened as their habitats in the wild are destroyed.

CAMOUFLAGE TECHNIQUES
Plants called 'living stones' (*Lithops*) are so well camouflaged in their surroundings that they are safe from animal grazing - until they flower.

Conservatory

THE SEASONALLY DRY ZONE is an enclosure watered sparingly in the winter. It contains plants from the East African deserts and savannah, which have all adapted in different ways. For example, the baobab stores water in its thick trunk, while acacias shed leaves in the dry season to conserve water.

Nearly 20% of the world's population lives on the edge of desert, where drought and often inappropriate farming techniques bring danger to semi-arid ecosystems. Kew is learning more from its living collection, studying the use of plants as sustainable crops for food, fuel, fodder and medicine, and to help slow down the rate of desertification.

THE WET TROPICS ZONE

This large area is maintained with the high humidity typical of rainforests. The planting also reflects the lighting conditions; low at floor level where species such as the marantas, with their attractively patterned leaves, are flourishing. Also on show are the wild forms of many familiar house plants, such as the Swiss cheese plant, the African violet and begonias.

SPECIAL AREAS FOR PARTICULAR PLANTS

Carnivorous plants are found in cool, well-lit areas and, since they often grow on poor soils, need to supplement their diets by catching insects. The best known species, the Venus flytraps and pitcher plants, are in the east of the Conservatory.

GOLD, FRANKINCENSE AND MYRRH
Boswellia sacra is the source of frankincense from biblical times - a resin that comes from wounds in the tree which, when dried, had medicinal qualities as a general antiseptic. There is also a relative of myrrh at Kew, but sadly, the collection lacks gold.

Orchids are given two distinct zones to provide the growing conditions that different types require. A hot, steamy zone features tropical epiphytic or air-rooting varieties with spectacular showy flowers and specific adaptations to an aerial environment in the rainforest canopy. The cooler orchid zone suits species with their roots in the earth of tropical mountainous regions.

Ferns, too, come from both tropical and temperate regions and there are two separate areas for ferns which reflect their different needs.

The Princess of Wales

The design challenge for this conservatory was both technical and aesthetic. Technically, it had to replace no fewer than 26 elderly glasshouses with one advanced and sophisticated building. Aesthetically, the site was exceptionally sensitive, being close to existing works of the past 'greats' - Burton's Palm House and Chambers' Orangery.

The keynote for the design was the highest possible energy efficiency allied to the lowest possible maintenance costs. It's a simple fact of life that the high humidity and high temperature needed to support life for tropical plants mean a slow death for inappropriate buildings.

With its stepped and angled glass construction, without sidewalls and with most of its space below ground, the conservatory is a most effective collector

of solar energy. The volume is relatively low in relation to its floor area so that temperatures within the individual zones may be altered quickly.

Above all, the Princess of Wales Conservatory looks stunning. The old adage that 'form follows function' is as perfectly demonstrated today as it was over 160 years ago in Burton's day. It looks absolutely in place in its surroundings, even in such imposing company. It is beautifully landscaped, too, blending into the Rock and Woodland Gardens to the east and south and surrounded by the mature trees of the Arboretum to the north and west.

Ten different environments, ranging from the extreme temperatures of desert to the suffocatingly moist heat of mangrove swamps, are controlled by computer to provide different levels of heat, humidity and light. The technology involved is quite

awesome. Sensors on walls and in beds report exact environmental conditions to the computer, which commands heat to flow, ventilation to open, or mists to spray to increase humidity. It is almost as if the building itself is a living organism.

The underground boiler room is a temple to today's technology. Also beneath the conservatory there are two 227,000 litre (50,000 gallon) storage tanks for rainwater collected from the roof slopes and used, after filtration and ultraviolet treatment, for irrigation and replenishment of the ponds.

Among its many awards the Princess of Wales Conservatory was presented with the Institution of Structural Engineers special award in 1987 and the highly prestigious Europa Nostra Award for Conservation in 1989.

Designed for an estimated life of 100 years, it is fascinating to wonder if the Princess of Wales Conservatory will be as compelling in its future old age as Burton's creations are today. Judging from the crowds, the comments and the awards, the omens seem good.

Conservatory

The Davies Alpine House

THE 'MOUNTAIN PEAK' SHAPE OF KEW'S NEWEST GLASSHOUSE REFLECTS THE HABITAT INSIDE

Built over the course of 2005, and opened in early 2006, this sophisticated new house is specifically designed to provide the right conditions for alpine plants, keeping them as dry and cool as in their natural habitat.

ABOUT THE DESIGN

The distinctive shape of the Davies Alpine House allows cool air to be drawn in around its sides while warm air escapes through vents high in the roof. Shading helps to keep the air cool, while the 12 mm thick laminated glass with its low iron content allows 90% of light through. The beds have plenty of well-drained soil with sub-surface drainage, and are landscaped with the Sussex sandstone recycled from the old alpine house. In 2006 the Davies Alpine House was awarded a prestigious RIBA award for architectural design.

SHADING THE PLANTS

Working with the sail-making industry made it possible to design a shading system that follows the line of the glasshouse. Visually pleasing, it is best described as four huge Japanese fans, hinged at the middle of each wall. A computerised system of ropes, pulleys and winches automatically opens the fans when shading is required. Each side of the glasshouse has independent control so that only the 'sunny side' is shaded.

COOLING THE HOUSE

In addition to the air vents, there is a cooling system using a basement chamber as a 'heat sink'. During the day (in the summer) air is drawn into the system from outside and passed through a labyrinth of walls where it is cooled before entering the glasshouse. The objective is to 'flood' the plants with cooled air. At night, when the ambient temperatures are cooler, the air flow continues into the labyrinth, while any warmth in the basement exhausts to the outside. In this way the labyrinth is cooled ready to supply cool air to the glasshouse the following day. The house stays within a temperature range of 0° to 28° C.

SMALL BUT PERFECTLY FORMED

From wild, open mountainsides, meadows and rock crevices, to Arctic shores and even the Mediterranean - alpines grow in a huge variety of places.

To define alpine plants, it is easier to think of the conditions that they need to grow: true alpines have short wet summers and long dry winters. They flower and set seed in a small window of opportunity between these two extremes – often as soon as water becomes available.

Most alpines grow on mountain slopes around the world and are highly adapted to their extreme environment. They survive winters under a protective layer of snow, with deep roots in well-drained soil. The effects of wind and high UV levels are often counteracted by staying compact, with small, hairy or succulent leaves that help stop them drying out.

Alpine gardeners often also grow beautiful bulbs from the Mediterranean. These are not true alpines as they need a long dry summer and a wet winter, but are a wonderfully colourful complement to the alpine garden.

The superb plants in the Davies Alpine House are just a fraction of the collection held behind the scenes. The display is regularly changed as plants come into flower. Over the year, hundreds of different plants enjoy the spotlight and are then taken back into the nursery once they have flowered.

SUNBURN FOR ALPINES?

In strong bright sunlight in high clear mountain air, humans are very susceptible to sunburn. It's exactly the same for plants and the red pigment on some alpines protects them against the effects of UV radiation and helps to absorb heat. The energy gained by these plants, like the *Sempervivum tectorum*, keeps them up to 12°C warmer than air temperature.

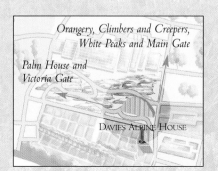

Orangery, Climbers and Creepers, White Peaks and Main Gate

Palm House and Victoria Gate

DAVIES ALPINE HOUSE

WINTER PROTECTION

Alpines spend their winters dry and dormant, protected by a thick insulating layer of snow. Melting snow gives them moisture for growth, beginning a short growing season, with intense light during spring and a relatively cool summer. Some plants survive the cold by insulating themselves with dense, fur-like hairs. Others trap daytime heat in their swollen stems and conserve it for freezing nights. Some protect themselves against strong winds by adopting a ground-hugging posture. The saxifrages are good examples - *Saxifraga cochlearis* also has small overlapping leaves to conserve moisture and a white waxy secretion to reflect strong sunlight.

The Evolution House

The Evolution House is a fascinating journey through millions of years of plant evolution showing not only the changes in plants, but how plants themselves changed the world. The story begins 3.8 billion years ago with the lifeless, barren landscape that existed then. The first life forms, 3.5 billion years ago, were *cyanobacteria* - living mats on the surface of structures called *stromatolites.*

Land plants developed in the Silurian period. *Cooksonia* was one of the first vascular plants - with conducting channels or 'veins' for carrying sap - and there is fossil evidence for this from about 410 million years ago. It was around 6 cm tall and grew near water.

The Evolution House concentrates on three major periods in plant evolution - the Silurian, Carboniferous and Cretaceous periods - and includes a coal swamp showing the giant clubmosses and horsetails from 300 million years ago. Cycads appeared 200 million years ago followed by conifers and the flowering plants, which dominate the plant kingdom today.

Among the species featured here are the world's largest horsetail (*Equisetum giganteum*) and the high-climbing fern *Lygodium*, whose fronds can grow to over 30 m (97.5 ft) long.

COAL - BLACK MAGIC?

In the Carboniferous period of geological time, clubmosses, tree ferns and giant horsetails flourished in warm damp conditions in extensive areas of wetlands. When these plants died, they fell into the airless morass where, instead of rotting away, they created coal swamps.

Coal is formed when a large concentration of plant debris is buried and subjected to moderately high temperatures and pressures. For example, geological events such as earthquakes or volcanic activity might bury the plants under many metres of rock or water. Under these conditions, various chemicals are driven out, the structure becomes fossilised and eventually turns to coal. The process started some 360 million years ago and went through stages, forming first peat and then lignite, or soft brown coal, before hard, black coal was created.

'Steam coal', still used to fire trains and other industrial steam engines in many parts of the world, is one of the coals which sometimes split to reveal well-preserved fossil plants. The models of giant clubmosses on display here are based on such fossils found in coal.

Model of giant woodlouse (*Arthropleura*) that would have lived at the time of the coal swamps.

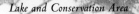

Lake and Conservation Area

Japanese Gateway
and Pagoda   Palm House and Main Gate

CAVE

CAVE ENTRANCE   ROCK FACE

CRETACEOUS
145-65 million years ago (MYA)

ROCKY OUTCROP

CARBONIFEROUS
360-290 MYA

COAL SWAMP

SILURIAN
435-410 MYA

ROCK BRIDGE

PRECAMBRIAN
More than 570 MYA

EVOLUTION HOUSE
ENTRANCE

TEMPERATE HOUSE

EVOLUTIONARY GROUPS

Just like humans and apes, many plants are related from a common ancestor in prehistory. There are many similarities between *Senecio* and *Rudbeckia*, for example, and their common origins are recognised by placing both in the same Compositae family. Living descendants of some of the examples in the Evolution House are cycads in the Palm House and Temperate House; conifers in the Arboretum; and ferns in the Princess of Wales Conservatory. More family resemblances in the world of flowering plants may be seen in the Order Beds by the Davies Alpine House, and in the neighbouring Grass Garden.

The Waterlily House

HOT AND HUMID – IDEAL CONDITIONS FOR SOME SPECTACULAR PLANTS

The Waterlily House is next to the Palm House. It is another of Kew's classic listed buildings, again with ironwork by Richard Turner. Built in 1852, it was then the widest single span glasshouse in the world, designed specifically to house the huge attraction of the age, the giant Amazonian waterlily.

Sadly, the huge plant never did well and in 1866, the house was converted into an Economic Plant House for medicinal and culinary plants. However, in 1991, it was converted back to its original use and today, it is the hottest and most humid environment at Kew and contains, as well as waterlilies, other very interesting plants.

In summer, the *Nymphaea* waterlilies and a giant *Victoria cruziana* put on a beautifully serene display. Sacred lotus and papyrus both thrive in these hot, humid conditions. In the corner beds, there are plants of economic importance such as rice, taro, bananas, manioc, sugar cane and lemon grass. High up, there are some spectacular gourds - fruits of some members of the cucumber family, such as loofahs, hedgehog gourds and wax gourds. The Waterlily House is closed in winter.

REPRODUCTION WITH BED AND BREAKFAST INCLUDED

When working in the Amazon forests, one of Kew's previous directors, Sir Ghillean Prance, discovered how giant waterlilies are pollinated. The flowers open at dusk, when their scent attracts beetles, who spend all night feeding on nectar as the flower closes around them. The next evening, when the flower reopens, the beetles - now covered in pollen - move on to another bloom, pollinating it in return for their 'bed and breakfast'.

Lake, Bamboo Garden, Japanese Minka and Rhododendron Dell

Lilac Garden, White Peaks, Climbers and Creepers and Brentford Gate

WATERLILY HOUSE

Princess of Wales Conservatory

The Bonsai House

SMALL AND PERFECTLY FORMED

At the northern end of the Order Beds - furthest from the Victoria Plaza -
there is a small glasshouse with a wonderful collection of bonsai trees. Bonsai
originated in China, but were taken up in Japan more than a thousand years ago.
The Japanese refined and developed Bonsai techniques and evolved the art form it
is today. 'Bonsai' means 'tree in shallow pot' and Bonsai masters aim to emphasise
the character of the species, bringing out the specific features that, in normal
growth, give it charm and grace. Kew's Bonsai are shown only a few at a time,
so they can be appreciated individually; but the display is constantly changing.

The Bonsai House has been completely refurbished. It once housed
Kew's alpine plants until a second Alpine House was built in
1981. This, too, has also been replaced; the renowned
Kew collection of alpines is now housed in the new,
strikingly innovative Davies Alpine House in the
Rock Garden.

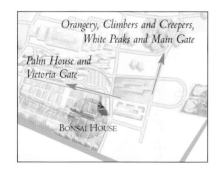

*Orangery, Climbers and Creepers,
White Peaks and Main Gate*

*Palm House and
Victoria Gate*

BONSAI HOUSE

47

Kew's special gardens

DUKE'S GARDEN · GRASS GARDEN · AQUATIC GARDEN · SECLUDED GARDEN

One of Kew's greatest charms is the sense of discovery. Turn a corner, follow a different path, and there's a stunning garden showing not just the beauty of the plants but the relationships between them. New knowledge and deeper understanding are great joys too, and very much a part of the pleasure of Kew.

DUKE'S GARDEN

The walled Duke's Garden is planted more for pleasure than science. Familiar shrubs and herbaceous perennials around large lawns give floral interest throughout the year. This tranquil walled area is home to the *Exotic Border*, a place for experiments with plant hardiness. Cannas and gingers are among the tender plants left out for the winter, and 'survival of the fittest' will determine how the border evolves. The *Gravel Garden*, sponsored by Thames Water, shows a wide range of attractive plants needing less water than traditional English garden choices.

Outside the walls, the *Duchess Border* carries the Lavender Species Collection, with both well-known garden lavenders and a number of tender species, rarely grown in the UK. Various Mediterranean specimens are trialled here to determine their hardiness in the south of England.

GRASS
GARDEN

AQUATIC
GARDEN

DUKE'S
GARDEN

GRASS GARDEN

Designed in 1982, the Grass Garden contains 550 species and the count is rising. Grasses (Gramineae, also known as Poaceae) are some of the most economically important plants, supplying food directly as cereals and indirectly as cattle fodder; they are the basis of many alcoholic drinks; and are used in building - straw thatch and bamboo - while sorghum and sugar cane (grown in the Waterlily House and Palm House) are used to produce petrol substitutes.

Visit in early summer for temperate annual grasses and cereals, and autumn for the perennials and their seed heads. Other display beds have British native grasses, tropical and temperate cereals, and specimen lawns showing different mixes of grass seed for different purposes. Of all the ornamentals, *Miscanthus sinensis* is notable for both its history - now a widely used architectural plant but having been grown from a single batch of seed - and its future, being investigated at Kew as an alternative and renewable fuel source.

AQUATIC GARDEN

Best visited in the height of summer when the waterlilies and aquatic plants are in full flower, the Aquatic Garden was opened in 1909, complete with hot water pipes to give the plants an early start. Today, some 40 varieties of waterlilies and more than 70 other plants such as sedges and rushes, are grown in containers in the central and corner tanks at ambient water temperatures, while the long side tanks hold other aquatics. As well as the beautiful waterlilies, the flowering rush and the bog bean are fine specimens.

SECLUDED GARDEN

This relaxed and intimate cottage-style garden is designed to appeal to all the senses, illustrated with poems on sight, scent, hearing and touch. Behind earth mounds, pleached limes form a hedge on tall stems to circle a spiral fountain. A stream is bordered with waterside plantings and scented flowers.

Even bad days are interesting when rain splashes off the giant leaves of *Gunnera tinctoria* and wind rustles through a tunnel of whispering bamboos. The planting here most closely resembles private gardens, with apple, pear and quince trees, roses, hostas, lilies, irises and salvias.

SECLUDED
GARDEN

Kew's special gardens

WOODLAND GARDEN · ROCK GARDEN · ORDER BEDS · ROSE PERGOLA

"I am at Kew, profiting by the exceptional summer to throw myself into 'plein air' studies in this wonderful garden of Kew. Oh! My dear friend, what trees! what lawns! what undulations of the ground!"

Camille Pissarro, visiting London, 1892

Some of the 'undulations of the ground' that so delighted Pissarro may well have started with something as prosaic as a gravel pit. Some 'ups and downs' are perfectly matched - the mound on which the Temperate House stands is spoil dug excavating the Lake. One thing is certain: that his delight with the Gardens, so enthusiastically expressed more than a century ago, is an experience shared by Kew's visitors today.

WOODLAND GARDEN

Here, a charming garden exactly replicates nature's design, since it is in three layers. High up, a deciduous tree canopy of oaks and birches shades the middle layer of deciduous shrubs such as maples and rhododendrons which, in their turn, protect the ground-cover layer, including hellebores, primulas, Himalayan blue poppies and North American trilliums.

Hellebores and peonies are tough hardy herbaceous and shrubby perennials. Visitors envious of the splendid hellebores should be reminded of the secret for success, which is never to disturb the roots. A few years ago, a panellist on BBC Radio 4's *Gardeners' Question Time* told the story of how his mother moved her much-loved 'Christmas Rose'. He told her that this was the last thing she should have done to a hellebore. So she moved it back again!

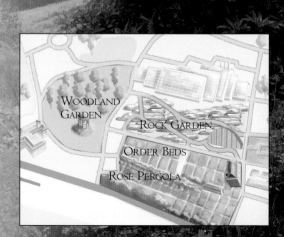

WOODLAND GARDEN

ROCK GARDEN

ORDER BEDS

ROSE PERGOLA

ROCK GARDEN

The first rock garden was designed in 1882 to look like a Pyrennean mountain valley. It is now built from sandstone, which retains more water than the original limestone, and has been redesigned to include a central bog garden and cascade. This makes for a wider variety of environments, in which alpines, Mediterranean plants and woodland and moisture-loving plants flourish. The alpines have a well-drained soil and a grit mulch to prevent water splashing on to their leaves; Mediterranean plants (from five regions of the world with similar climates) are in the sunniest spots, again with well-drained soil; while the woodland plants enjoy the shade and damp conditions created by the water features.

There are six global regions represented in Kew's attractive Rock Garden. Visitors used to domestic rock gardens are intrigued both by the size of the display and being able to walk through it, by the waterfalls and gullies; and by the truly extraordinary variety of the world's plants on show here.

ORDER BEDS

The Order Beds are a 'living library' of flowering plants for students of botany and horticulture, as they are arranged systematically in family groups, so they can be easily located for study. Here, Kew's own students plant vegetables as part of their course. There is more about *taxonomy*, which is the grouping of plants and the science of understanding the relationship between them, on pages 78 & 79. Science apart, the Order Beds are stunning to see in full summer bloom.

ROSE PERGOLA

A pergola covered with a variety of climbing roses is a deservedly popular feature in the central path of the Order Beds, while the surrounding walls support many fine climbing plants, such as *Actinidia kolomikta* with its striking pink, red and white spring foliage.

51

Kew's special gardens

QUEEN'S GARDEN · BEE GARDEN · WINTER GARDEN

"I never had any other Desire so Strong and so like to Covetousness, as that one which I have had always, that I might be Master at last of a small House and large Garden … and there to dedicate the Remainder of my Life to the Culture of them, and study of nature."

Abraham Cowley (1618-67) English poet and essayist

QUEEN'S GARDEN

Kew Palace is the oldest building at Kew (see p 80) and behind it, there is a charming 17th century style garden. The word 'style' is used advisedly because despite its historically impeccable appearance, it was conceived in 1959 by Sir George Taylor, then Director of the Royal Botanic Gardens, and officially opened by H.M. Queen Elizabeth II ten years later.

One element is a parterre enclosed in box hedges and planted with lavender, cotton lavender, purple sage, variegated iris, lady's mantle and Russian sage among others. Standing in the pond to the rear of the parterre is a copy of Verocchio's 'Boy with a Dolphin', the original of which is in Florence's Palazzo Vecchio. Other attractive features are the sunken Nosegay Garden, full of sweet-smelling and useful flowers and a traditional vegetable garden.

The plants in the Queen's Garden are exclusively those grown in Britain before and during the 17th century. Their labelling differs from Kew's norm, since they include not only today's botanical name and family, but also:-

- the common name in the 17th century
- a virtue, or quotation from a herbal (medicinal plant book) and
- the author's name and date of publication.

There is a great deal to see in the Queen's Garden - a wrought iron pillar from Hampton Court Palace, a chamomile chair, intricately plaited laburnums forming an arch, pleached hornbeams making a hedge on tall stems, a mound covered in clipped box, with a gazebo and a fine Judas tree.

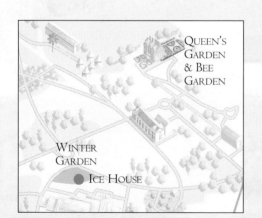

QUEEN'S GARDEN & BEE GARDEN

WINTER GARDEN

ICE HOUSE

BEE GARDEN

No English garden is complete without its bees and Kew is no exception. Developed in 1993, the Bee Garden has three styles of beehive, from simple logs and basketware skeps, to modern wooden hives that grow taller as more 'supers' (extra layers) are inserted as the honeycombs form inside them. Two colonies had new queens flown in from New Zealand where bees are traditionally quite docile. The Bee Garden is full of flowers popular with the bees and shows the important interaction between plants and insects. Without the pollinators, plants would not set seed and our food supplies would dwindle rapidly.

WINTER GARDEN

The Winter Garden around the Ice House (see p 71) is laid out for real interest on grey days. Both sight and scent come into play in this aesthetically satisfying garden, where specimens are set off against an evergreen backdrop, perfect for those which flower on bare wood. Here, *Mahonia x media* 'Winter Sun', winter box and the chocolate-scented *Azara microphylla* surround wintersweets, viburnums, flowering quince, cornelian cherries, witch hazel and willows with their yellow catkins. There are bulbs, too, planted under the woody specimens, with winter aconite, snowdrops and windflowers left to naturalise. Conservationists will note *Abeliophyllum distichum* which is endangered in its native Korea.

53

Kew's special gardens

LILAC GARDEN · ROSE GARDEN · RHODODENDRON DELL
AZALEA GARDEN · MAGNOLIA COLLECTION · BAMBOO GARDEN

*"Go to Kew in lilac-time ... and you shall wander
hand in hand with love in summer's wonderland."*
From 'The Barrel-Organ' by Alfred Noyes (1880-1958)

LILAC GARDEN

The poet Alfred Noyes was enchanted by lilacs at Kew and rightly so, as they are among the most elegant and colourful of all early-flowering shrubs. Several lilacs, including the Chinese species *Syringa julianae* and *Syringa sweginzowii*, came into cultivation through Kew's lilac collection. Today's Lilac Garden is the result of a replanting completed in 1997. Most lilacs are cultivated from eastern Europe's *Syringa vulgaris*, but lilacs can also be found through the Himalayas to the mountains of China where the greatest diversity occurs.

In Kew's Lilac Garden, best viewed in early June, there are 100 specimens in ten beds arranged according to their cultivation and breeding history. As well as *Syringa* species, there are 'Chinese' lilacs based on the hybrid *Syringa x chinensis*; 'Hyacinthiflora' hybrids from a cross between *Syringa vulgaris* and the early flowering Chinese species *Syringa oblata*; and the late flowering Villosae lilacs from Canada.

ROSE GARDEN

This is a major draw for visitors, and is in full glorious bloom from June to August. Cluster-flowered and large-flowered roses are arranged by colour; red nearer the Palm House, contrasting against the white building; shading out to lighter colours at the perimeter, where white and yellow roses are set against the green hedges and vistas.

Each of the 54 beds contains a different cultivar of rose, all available through British rose growers. Ten of the beds illustrate the hybridisation of roses through the centuries.

The semicircular holly hedge and sunken areas are historically important, being the remains of an 1845 design by William Nesfield.

ROSE GARDEN   LILAC GARDEN

RHODODENDRON DELL

Rhododendrons are one of the largest and showiest groups of flowering shrubs, with great variation in size, habit and form. There are over 700 specimens planted in the Dell, with some unique hybrids found only here. This is not a natural valley, but a creation of 'Capability' Brown, who carved what he called the 'Hollow Walk' in 1773.

In the 1850s, Sir Joseph Hooker started the rhododendron collection with specimens gathered on his Himalayan expeditions. The oldest specimen is *Rhododendron campanulatum*; the most highly scented are *Rhododendron kewense* 'King George' and *Rhododendron loderi*. 'Cunningham's White' puts on a brilliant show each spring; it thrives in the local conditions, shaded by trees and kept humid by the Thames nearby. Flowering continues from November to August, with most at their best in late May.

AZALEA GARDEN

Anyone looking for azaleas to grow for their blaze of late spring colour will be spoilt for choice at Kew. Twelve different groups of azalea hybrids have appeared since the very first Ghent hybrids in the 1820s and the garden's two circles of beds arrange the plants in date order, leading up to today's varieties from eastern America and Holland. All the species planted in the Azalea Walk leading into the garden belong to a group of deciduous azaleas from North America and Japan.

MAGNOLIA COLLECTION

The first exotic magnolia was introduced to this country in the seventeenth century. Many of these deciduous and evergreen shrubs and trees with their spectacular pink, cream or white flowers, do very well here, as can be seen from March to June in a superb planting near the rhododendrons and azaleas.

BAMBOO GARDEN & JAPANESE MINKA

A year-round pleasure to visit, the Bamboo Garden makes the most of bamboos' variety of forms, stem colours and leaf shapes. Bamboos are woody grasses ranging from giant poles, through wispy variegated species, to fountains of leaves from the pendulous varieties. Some specimens hardy in the UK can reach 3 m (10 ft) high, others can be mown as lawns! Bamboos grow wild on every continent except Europe and are used for everything from food to building materials.

In the centre of the Bamboo Garden there is a traditional Japanese Minka house. Its story and how it came to Kew is told on page 69.

The Arboretum

What is it about woodlands and trees? Mankind's fascination with them must be rooted in prehistory, when they provided shelter and fuel, weapons and tools. Exotic trees have been collected at Kew since its earliest days, but it was during the stewardship of Kew's first Director Sir William Hooker that today's Arboretum took shape.

WOODLAND GLADE

The Woodland Glade is notable for its giant redwoods and other majestic conifers, underplanted with shrubs for summer and late autumn colour. There's a sense of great calm by the charming Waterlily Pond with its profusion of aquatic life as well as plants.

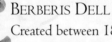

BERBERIS DELL

Created between 1869 and 1875, this was once a gravel pit and is Kew's third-largest excavation after the Lake and Rhododendron Dell. Its rather secluded character and large collection of *Berberis* and *Mahonia* make it well worth a visit. Most of the berberis family are fully hardy in Britain and grow well in most soils. Flowers in shades of yellow and orange bloom from early spring to late summer, while the autumn fruit can be red, black or blue. A wonderful Emily Young sculpture makes a striking focal point in the Dell.

HIGHLY COLOURED

Autumn colour reaches its peak with the vibrant leaves of the maples between the Temperate House and the Marianne North Gallery; while still more colour comes from the berries and foliage on various trees of the Rosaceae (rose) family; whitebeams, rowans, hawthorns and crab apples.

CHERRY WALK

Stretching from the Rose Garden behind the Palm House to the Pagoda, Cherry Walk is a superb collection of Japanese ornamental cherry trees. Replanting of the original 1935 walk was completed in 1996 and now, 22 trees in 11 matched pairs of different cultivars make a spectacular show in spring. The charming and expressive Japanese names for the cherries include *Imose* (sweetheart), *Tai Haku* (big white flowers) and *Taki-nioi* (fragrant waterfall).

THE LAKE AND SACKLER CROSSING

There have always been water features at Kew, and the Lake, which despite appearances is artificial, was started with the present Pinetum and extended when the Temperate House was built. Moisture-loving trees and shrubs are planted all round, while the islands are planted with nyssas for their autumn colour, which reflects with spectacular effect in the water, together with several rare Wollemi pines. In 2006 the Lake was renovated and the Sackler Crossing was built, elegantly spanning the water and opening up new vistas. Designed in a sinuous line to reflect its natural setting, the granite and bronze Crossing gives the effect of walking on water.

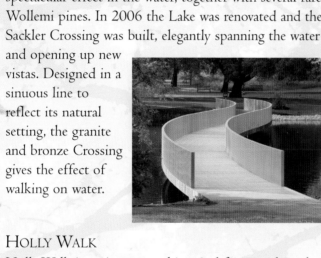

HOLLY WALK

Holly Walk is an important historical feature, planted along what used to be Love Lane on the main public right of way from Richmond to the Brentford ferry.

Originally laid out in 1874 by Sir Joseph Hooker as part of a series of vistas, most of the hollies are original - over 135 years old - and many are now large trees. At over 1030 m (3390 ft) in length, Holly Walk is the largest, most comprehensive collection of mature hollies in cultivation. The picturesque berries, which vary from red to black or white depending on the species, are at their best in November and December, but continue through to the spring.

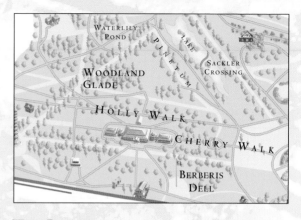

THE PINETUM

The first Pinetum at Kew was planted nearly 250 years ago near the Orangery. The second was created near the Palm House and some of those conifers can still be seen. Today's Pinetum, at the southern end of the Gardens, was started in 1870 by Sir Joseph Hooker, when the trees were laid out with related species grouped together.

In 2008 a permanent Treetop Walkway and 'Rhizotron' underground experience will open to the public in the Arboretum offering visitors two brand new ways to look at trees.

Kew's Heritage Trees

The Arboretum contains specimen trees dating from as far back as the early 18th century to newly planted rarities from recent Kew plant collecting expeditions. Each tree is a piece of history in an ever-changing landscape, representing a great historic tree collection over generations.

SWEET CHESTNUT, *CASTANEA SATIVA* EARLY 18TH CENTURY

This ancient tree is probably the oldest in the Gardens. The several sweet chestnuts here are thought to be the remnants of the woodland planting that lined Love Lane and separated the Kew and Richmond estates. Pollarding at various stages of their lives has given them great character and form.

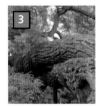

'THE OLD LIONS' 1762

'The Old Lions' are some of the few remaining trees known to have been planted in the original garden of Princess Augusta around 1762. They are the maidenhair tree (*Ginkgo biloba* - 2), Japanese pagoda tree (*Styphnolobium japonicum* - 3) by the Princess of Wales Conservatory; the Oriental plane (*Platanus orientalis* - 4) near the Orangery; the false acacia (*Robinia pseudoacacia* - 5) on the Orangery lawn; and the *Zelkova carpinifolia* (6) in the Herbarium paddock.

The maidenhair tree, a male, was one of the first to be planted in Britain. No-one knew how hardy it was, so it was trained like a fruit tree against the wall of the Great Stove glasshouse for shelter. When this was demolished in 1861, it was left standing alone. In 2002 it rightly became one of the 50 'Great British Trees' in a scheme run by the Tree Council to celebrate the Queen's Golden Jubilee.

THE LUCOMBE OAK, *QUERCUS X HISPANICA* 'LUCOMBEANA' 1773

The first Lucombe Oak was created in 1762 in the Exeter nursery of a Mr Lucombe, as a cross between turkey oak and cork oak (*Q. cerris* x *Q. suber*). Kew's specimen was planted around 1773, but was moved 20 m (65.5 ft) in 1846 to make way for the Nesfield's Syon Vista (see pp 60 & 61).

TURNER'S OAK, *QUERCUS X TURNERI* 1798

This semi-deciduous oak was bred in a Mr Turner's Essex nursery in the late 18th century, as a cross between English oak and holm oak (*Q. robur* x *Q. ilex*). Over the years the earth around the roots became compacted and the tree had begun to suffer. But during the Great Storm in October 1987, the whole root plate lifted, loosened and settled back in the ground. This seemed to rejuvenate it and was one of the factors that prompted the ongoing decompaction programme for the Arboretum's mature trees.

CORSICAN PINE, *PINUS NIGRA SUBSP. LARICIO* 1814

This large Corsican pine is sited where the first Pinetum at Kew was systematically planted following the Linnaean system. Early in the 20th century, a light aircraft crashed into it, taking out the top of the tree. Since then it has been struck by lightning at least twice, the last time in 1992, leaving only scars on the trunk as a reminder.

TULIP TREES, *Liriodendron tulipifera* & *Liriodendron chinense* 2001

The North American species of tulip tree, *Liriodendron tulipifera*, was introduced to Britain in 1688. In the Azalea Garden, a specimen planted in the 1770s still flourishes. The Chinese tulip tree, *Liriodendron chinense*, a far superior species, was introduced by Ernest Wilson in 1901, but is now very rare. Saplings grown from seeds collected during a recent expedition to China were planted in 2001 to recreate the Arboretum's old 'Tulip Tree Avenue'.

CHESTNUT-LEAVED OAK, *Quercus castaneifolia* 1846

The chestnut-leaved oak was introduced to Britain from the Caucasus and Iran as seed in 1843. The specimen behind the Waterlily House was Kew's first, planted in 1846. Today, at over 30 m (98.4 ft) in both height and spread, it is unrivalled as the biggest and finest specimen of its type in the world.

INDIAN HORSE CHESTNUT, *Aesculus indica* 'SYDNEY PEARCE' 1935

This fine tree, named after Sydney Pearce, the Arboretum's Assistant Curator in 1935, is a particularly good flowering form. The three mature specimens that can be seen on the Little Broad Walk were transplanted there from a collection near the Pavilion restaurant when they were seedlings. Another three trees have since been added to create a colourful avenue from the Main Gate.

STONE PINE, *Pinus pinea* 1846

Stone pines have been cultivated in Britain and continental Europe for some 2000 years for pine nuts - the seeds from their cones. However, Kew's first planting was not until 1846 and along with many other newly-introduced pines, this specimen was kept in a pot for many years, for lack of space in the original arboretum. In its pot, the stone pine became bonsaied, but once planted out when the arboretum expanded, it grew away to form the unusual shape it has today. Another name for the stone pine is 'umbrella' pine due to its parasol appearance when grown on a single trunk.

The Broad Walk and

The development of Kew Gardens can be divided into two main periods, before and after 1845, when the Gardens expanded to some 110 hectares (274 acres). It was then that the vision of Sir William Hooker, the first full-time Director; Decimus Burton, the supervising architect; and William Nesfield, the principal landscape designer, shaped the grand glasshouses and great sweeps of planting and landscaping so familiar and so admired today.

It was not always harmonious. Nesfield and Hooker had differences where design confronted taxonomy. Nesfield's creativity was stifled, he felt, by Hooker's insistence on grouping trees by their botanical relationships, not by size, colour and form. Hooker, on his part, said that highly ornamental parterres were not consistent with the nature of the Arboretum. However, the gaining of more land satisfied their needs.

By autumn 1845, Nesfield's plan for the entire gardens included the Broad Walk, lined with deodar cedars interspersed with flower beds; the Pond to the east of the Palm House, which had new parterres either side of it; the Pinetum and enlarged Arboretum; and new walks, vistas and paths.

The vistas and avenues are Nesfield's indelible signature on the Kew landscape. In a 'goose foot' pattern radiating from the Palm House, Syon Vista was a wide gravel-laid walk stretching 1,200 m (3,937 ft) towards the Thames; Pagoda Vista was a handsome grassed walk some 850 m (2,800 ft) long; while the third, short, vista fanned from the northwest corner of the Palm House and focused on a single 18th century cedar of Lebanon. These vistas and the parterres demonstrated the importance of Burton's Palm House to Kew and although many of the original formal beds have long since vanished, the impact of the grand plan remains to this day.

three great vistas

RESTORING NESFIELD'S BROAD WALK

Kew has embarked upon a plan to restore the grandeur of Nesfield's original planting design on the once-majestic Broad Walk. Sixteen semi-mature Atlantic cedars now take the place of Nesfield's deodar cedars. The Atlantic cedar looks very similar to the deodar, but grows better in the dry climate of London.

Moving plants can be tricky. Moving trees needs even more care and enormous power. An excellent tree transplanter, capable of carrying trees up to 20 m (66 ft) in height, had been designed by William Barron. Kew bought one in 1866 and used it to great effect, one winter transplanting 60 trees weighing up to seven tonnes each.

To move a tree, a deep trench had to be dug 2 m (6 ft 6 ins) away from, and all round the trunk. The transplanter's side beams were removed and the machine reassembled round the tree. A cradle was placed under the root ball and a system of ropes, pulleys and winches heaved the tree, upright, out of the ground. A team of horses then dragged the entire contraption to a prepared hole for replanting.

A fully restored Barron's Tree Transplanter being used to mark the start of the Broad Walk restoration in early November 2000.

Conservation and Wildlife

While Kew is a world class botanic garden and an enormously popular place to visit, it also has a global reputation as a centre of excellence for plant research. Kew scientists undertake the basic plant identification and classification studies that underpin all other scientific and conservation activities related to those plants. There is a vast amount of scientific effort put into supporting conservation and sustainable use of plant resources in the UK and overseas (see pp 85-88). Conservation of habitat both at home and abroad is vital not only for plant species and their dependent wildlife to survive, but also for the survival of mankind itself.

CONSERVATION AREA

Kew supports an extraordinary diversity of British wildlife, much of which can be found in the Conservation Area around Queen Charlotte's Cottage (see p 81). Its 15 hectares (37 acres) are actively managed to encourage native flora and fauna.

It is mainly woodland, with many British trees represented, including oak, beech, holly and yew. In spring, the woodland floor is a much-admired carpet of bluebells, wild garlic and snowdrops. There are also some rare native trees, such as the Plymouth pear and the Bristol mountain ash. Very few examples of these are left in

the wild, in Devon and the Avon Gorge respectively, and here, they are fine examples of Kew's conservation efforts in saving many threatened species from extinction.

Dead trees are left standing or where they fall. Holes and crevices in old, dying and dead trees provide excellent feeding and nesting opportunities for birds and roosting places for Kew's many bats (see p 65). They are also essential for the myriad fungi to thrive and eventually break down into valuable nutrients for the soil. More species depend on dying or dead wood than on living trees.

Hazel coppicing takes place on a regular cycle - cutting the trunks of trees to ground level and letting shoots grow up from the stumps. Coppicing keeps woodland open, letting understorey plants and wildflowers gain a foothold. The young hazel shoots grow into long sturdy poles to use for plant supports and even woven hurdles elsewhere in the Gardens.

There are meadows and grassy rides, wetland, ponds and a small gravel pit, all

at Kew

helping to support native butterflies, dragonflies and other insects. Amphibians, too, find a home here. Many of these species are in decline both locally and nationally due to changes in farming and new building bringing about the loss of many of their habitats.

Larger mammals, such as foxes and badgers also thrive in the Conservation Area and for the many visitors wishing to learn more about how badgers live, there is a human-sized badger sett to explore.

Please note that to help conserve plant and animal species in the Conservation Area, we ask visitors to keep to the hard-surfaced paths, the boardwalk and the elevated viewing platform with its views across the gravel pit, pond and wildflower meadows.

WILDLIFE OBSERVATORY

Dragonflies and damselflies (see p 65), frogs and newts, and a myriad of water-creatures can all be found in this area.
A wide range of habitats, plants and wildlife can be seen from the Observatory, where the benefits of conserving habitats and building garden ponds are explained. The Dipping Pond has been created as an educational tool for visiting school groups.

THE BADGER SETT

Walk through the giant forked oak branch entrances and experience what a badger's home is really like. Food stores, sleeping chambers and nests are connected by a warren of tunnels. The tunnels are all at least a metre high, and one, 1.5 metres high, is suitable for wheelchairs.

THE STAG BEETLE LOGGERY

This is Britain's largest native beetle and males can grow up to 70 mm (2.76in) in length, including the 'antlers', which are elongated jaws. Larvae feed in rotting wood for up to seven years before emerging briefly - usually in May and June - as mating adults. The dramatic 'Loggery' near Queen Charlotte's Cottage is part of the London Biodiversity Action Plan for Stag Beetles. The Thames Valley is a hotspot for stag beetles and Kew is doing all it can to encourage them. There is information here on how to build a home garden loggery.

Walk on the wild side

WHEN WALKING THROUGH KEW GARDENS, IT IS VERY REWARDING TO LOOK AT MORE THAN THE PLANTS AND THE BUILDINGS. IN THE SKY, ON THE GROUND, IN AND UNDER THE WATER, THERE'S AN ABUNDANCE OF WILDLIFE

TUFTED DUCK

SHOVELLER

WIDGEON

MANDARIN DUCK

POCHARD

CORMORANT

BARNACLE GOOSE

GREYLAG GOOSE

BAR-HEADED GOOSE

EGYPTIAN GOOSE

GREAT CRESTED GREBE

PHEASANT

RING-NECKED PARAKEET

SHELDUCK

BIRDS

Human birders flocking to Kew have reliably listed no fewer than 128 bird species. Trees, grass and scattered shrubs attract typical birds of London's wilder inner suburbs, such as kestrels and sparrowhawks, woodpeckers, blackcaps, chiffchaffs, nuthatches and treecreepers. Blue, great and coal tits, robins, and friendly introduced pheasants are in the dells. There are herons and cormorants on the Thames, while within Kew, great crested grebes nest on the Lake and Palm House Pond, with an annual invasion of wild waterfowl, supplementing the ubiquitous Canada geese and resident ornamental birds.

BEES

Kew's Bee Garden (see p 53) has three types of beehive and the Kew website lists native bees and describes 'A Honey Bee's Year' at www.kew.org/conservation/

BUTTERFLIES

Lepidopterists like Kew. At least 23 species of butterfly belonging to five families have been recorded at Kew since 1980 - a high number considering the London location, but a reflection of the variety of plant life and habitats within the Gardens. Nectar plants are important for the butterflies, but food plants for their caterpillars are just as vital, together with proper habitat management to allow each species to complete its annual life cycle. The mowing regime of many areas at Kew is designed specifically to provide nectar for the adults and sufficient food to see the caterpillars through to their pupal forms. Kew's butterflies are present at some stage of their life in winter, too. Some hibernate as adults, others as caterpillars or pupae, while the remainder overwinter only as eggs. In summer, they divide their time between eating and trying to produce the next generation.

Small Copper

Common Darter

Banded Demoiselle

DRAGONS AND DAMSELS

The difference between a dragonfly and a damselfly is easy to see when they're at rest. Dragonflies keep their wings outstretched like small helicopters, while most damselflies fold theirs back along the length of their bodies. Dragonflies are also usually more sturdily made than slender damselflies. The flying stage is the very last part of their lives, concerned with reproduction.

Eggs are laid either into plant tissue (under or near water), or directly into the water. When they hatch, the nymph stage can last several years in some species. The carnivorous nymphs have fierce prey-seizing jaws on an extendible 'mask', a feature unique to dragonflies and damselflies. They feed on insects and other small creatures but some, like the Emperor dragonfly nymph, can take small fish.

Mature nymphs crawl up plant stems into the open air for their final moult, when the adult flying form emerges, waiting for wings to expand and legs and body to harden before searching for food and a mate.

Blue-tailed damselflies are seen from early June around the Lake, with the Common Blue at the Waterlily Pond from July. The dragonflies start flying in July with Brown and Migrant Hawkers and the Common Darter. Hawkers, Darters, Chasers, Skimmers - dragonflies and their close damselfly relatives are a fascinating study.

BATS

Kew's conservation policies encourage bats and of the 15 British native species, Kew is home to at least three, and probably more. In recent years, pipistrelle, noctule and Daubenton's bats have been recorded.

FUNGI

Autumn is an especially fruitful time for fungi. What appears above the ground, or from tree trunks and rotting vegetation as mushrooms, toadstools and brackets, is only the fruiting stage, and short-lived at that. The main body of a fungus lives within its source of food, which could be wood or other plant material, the soil, or carrion. Fungi are a separate group, neither plant nor animal, consisting of microscopic threads that form a network that expands out through, and feeds on, organic matter. Fungi are usually the primary decomposers in their habitats, releasing nutrients to the surroundings. About 80% of all plants grow in mutual association with fungi and many, from orchids to pine trees, will not grow without a partner fungus within their root systems - a mycorrhizal association. Scientists at Kew are studying fungi from around the world, investigating their practical uses as biological controls against pests such as insects and eelworms, with the objective of reducing the use of toxic pesticides.

There is far more to 'mushrooms' than meets the eye and an illustrated leaflet, 'Mushrooms & Toadstools', is normally available from the Friends' desk in the Victoria Plaza.

The Pagoda

ONE OF KEW'S FAMOUS FEATURES, THE PAGODA IS ONE OF 25 ORNAMENTAL BUILDINGS DESIGNED FOR KEW BY SIR WILLIAM CHAMBERS

There was a fashion for *chinoiserie* in English garden design in the mid 18th century, and Chambers was a keen advocate, using decorative buildings and intricate pathways as a reaction to the sweeping 'natural' lines of contemporaries such as 'Capability' Brown.

The Pagoda was completed in 1762 as a surprise for Princess Augusta. It was one of several Chinese buildings completed for the Kew estate by Chambers and added a final flourish to the end of the garden.

It was originally flanked by a Moorish Alhambra and a Turkish Mosque.

The ten-storey octagonal structure, constructed by local builder Solomon Brown, is 163 ft (50 m) high and was, at that time, the most accurate reconstruction of a Chinese building in Europe (although purists argue that pagodas should always have an odd number of floors). It tapers, with successive floors from the first to the topmost being 1 ft (30 cm) less in both diameter and height than the preceding one.

A COLOURFUL PAST

The original building was intended to be very colourful; Chambers wanted the roofs to be made of green and white varnished iron plates; the banisters to be a mix of blue, red and green; and with 80 sinuous gilded dragons to decorate the roof corners and a tall gold finial to top it all off. However the iron plates were replaced by slate and sadly, the dragons vanished 22 years later in 1784, during maintenance works. It is not known if Chambers did complete his colour scheme as it survived for so short a time that records are scarce.

Since then, the Pagoda has undergone no fewer than 24 colour scheme changes. In 1843, Decimus Burton wanted to restore the Pagoda to its former glory, but the cost then of £4,350 was considered too high a price to pay. It has been opened to the public only a handful of times in its history, most recently in 2006 for Kew's heritage year festival. It is hoped that the Pagoda can be restored and opened again in the not too distant future.

BOMBS AWAY!

Contemporaries of Chambers often wondered if such a tall building would remain standing, though it had been "built of very hard bricks". Its sturdy construction was proved in World War II when it survived a close call from a stick of German bombs exploding nearby. This was ironic, since at the time, holes had been made in each of its floors so that British bomb designers could drop models of their latest inventions from top to bottom to study their behaviour in flight.

SIR WILLIAM CHAMBERS (1723-1796)

Born to Scottish parents in Gothenburg, William Chambers went to live in China as an employee of the Swedish East India Company, a stay which sparked his interest in Chinese architecture. At the age of 26, he studied to become an architect and trained classically in Rome. In 1757 he became Princess Augusta's official architect and architectural tutor to her son, the future George III. He supervised the buildings at Augusta's five establishments and, over a five year period, designed 25 decorative buildings for Kew. Many, such as the Mosque, the Alhambra, the Palladian Bridge and Menagerie have disappeared, but the Orangery, the Ruined Arch, the Temple of Bellona and, above all, the Pagoda, are permanent reminders of his talents. His most solid achievement is Somerset House in London, recently restored to its original splendour. He was knighted in 1771 by King Gustav III of Sweden for a series of drawings of Kew Gardens, and George III allowed him to use the title in England.

Japan at Kew

Kew's Chokushi-Mon (the Gateway of the Imperial Messenger) is a four-fifths size replica of the Karamon of Nishi Hongan-ji in Kyoto, the ancient imperial capital of Japan. The replica was originally built for the Japan-British Exhibition, held in London in 1910, after which it was dismantled and rebuilt in Kew Gardens.

It is the finest example of a traditional Japanese building in Europe, built in the architectural style of the Momoyama (or Japanese rococo) period in the late 16th century, a time of peace, prosperity and flowering creativity.

Typically expressive, Chokushi-Mon shows finely carved woodwork depicting flowers and animals, with the most intricate panels portraying an ancient Chinese legend about the devotion of a pupil to his master.

RESTORATIONS

The Japanese wood-carver, Kumajiro Torii, carried out some detailed repair work in 1936 and 1957, but by 1988, the edifice was badly dilapidated. With generous support from Japan and elsewhere, a full restoration, combining traditional Japanese skills and modern techniques, was completed after a year's painstaking work in November 1995.

Today, Chokushi-Mon is seen in rather more than its original splendour, because as part of the restoration, the original lead-covered cedar-bark roof shingles were replaced with traditional, more dramatic, copper tiles.

THE JAPANESE LANDSCAPE

Around the Chokushi-Mon, the Japanese landscape is in three distinct areas, each depicting one of the many different aspects of Japanese gardens. Overall, it is a dry stone *kaiyu shiki* (stroll-around) garden in the Momoyama style.

The main entrance leads into a 'Garden of Peace', reminiscent of a tea garden (*roji*), a calming, tranquil place with stone baths and a gently dripping water basin.

The slope to the south of Chokushi-Mon is a 'Garden of Activity', symbolising the natural grandeur of waterfalls, hills and the sea. The third area, the 'Garden of Harmony', links the other two and represents the mountainous regions of Japan, using precisely positioned stones and rock outcrops, interplanted with a wide variety of plants of Japanese origin, many of which are well known in the West.

THE JAPANESE MINKA

Until the middle of the 20th century, many Japanese country people lived in wooden houses called minka. They had sturdy wooden earthquake-resistant frames for mud-plastered walls and thatched roofs - the equivalent of British wattle-and-daub cottages. In northern Japan, minkas had steep roofs and small windows to cope with the long snowy winters. In the hotter southern Japan they were low buildings with raised floors for good ventilation and resistance to typhoon damage. Minkas were often a workplace as well as a home.

Minkas became less popular in Japan during the latter half of the twentieth century, being considered inconvenient and uncomfortable. Many were demolished and replaced by modern houses lasting under 30 years. The constant building and replacing of newer houses created vast quantities of industrial waste, whereas minkas could be reconstructed or used as fuel. In 1997, the Japan Minka Re-use and Recycle Association (JMRA) was established to promote the benefits and conservation of minkas.

Kew's minka was originally a farmhouse built around 1900 in a suburb of Okazaki City, near the southern coast of central Japan. When the Yonezu family bought it in 1940, it was dismantled and moved to another part of the city. The family moved into it in 1945. After the death of Mrs Chiyoku Yonezu in 1993, JMRA acquired the house and donated it to the Royal Botanic Gardens Kew as part of the Japan 2001 Festival.

The dismantled wooden framework was shipped to Kew, where construction began on 7 May 2001. Experienced Japanese carpenters reinstated the intricate joints formed without the use of iron nails. A Japanese ceremony was held when the framework was completed on 21 May.

A team of British builders who had worked on the Globe Theatre in London built the mud wall panels. The roof was then thatched with Norfolk reeds and wheat straw. The Japanese Minka was completed in November 2001. Occasional displays of Japanese arts and artefacts add insight and interest to a visit.

The Japanese Minka in the Bamboo Garden (see p 55.)

Kew follies

When they were built, the decorative buildings at Kew represented all the fervent interest and excitement of knowledge newly gained in a widening world. China and the Islamic world were opening up. There was a revival of classicism through the 'Grand Tour' taken by 18th century gentry. This passion for the newly-discovered influenced the influential in the design and development of the Gardens at Kew.

The ornamental features at Kew are an eclectic collection and should be looked at in the context of their period. Past plans and illustrations show that so many have come and gone, that fervent thanks should be given for those that remain as treasured features of this World Heritage Site.

THE CHAMBERS COLLECTION

The Ruined Arch, Temple of Bellona, the Temple of Arethusa, and Temple of Aeolus were all designed and constructed by Sir William Chambers (see p 67). They are all Grade II listed buildings.

THE RUINED ARCH is a charming folly built in 1759 as a fashionable mock ruin.

THE TEMPLE OF BELLONA (1760) is named after the Roman goddess of war. Behind its Doric facade is a room decorated with plaques bearing the names of British and Hanoverian regiments which distinguished themselves in the Seven Years' War (1756-63). Despite its appearance it is fashioned from wood.

THE TEMPLE OF ARETHUSA (1758) is named after Arethusa, a nymph attendant of Diana the Huntress. Today this temple is the setting for Kew's war memorials and is adorned with wreaths every November 11th.

THE TEMPLE OF AEOLUS (1760-63) Aeolus was the mythical king of storms and the four winds, inventor of sails and a great astronomer. The temple was originally built of wood and once had a revolving seat to provide a panoramic view of the whole Kew estate. It was rebuilt in stone by Decimus Burton in 1845.

Ruined Arch

Temple of Bellona

Temple of Arethusa

Temple of Aeolus

King William's Temple

Ice House

DECIMUS BURTON
(1800-1881)

A brilliant architect from an early age, Burton was the son of a London builder. As the supervising architect at Kew, he designed the Palm House in conjunction with Richard Turner (see p 26), the Temperate House, Museum of Economic Botany, and Main Gate, worked with William Nesfield on the design for the Broad Walk (see p 60) and rebuilt the Temple of Aeolus in its present location.

KING WILLIAM'S TEMPLE

Built in 1837 by Sir Jeffry Wyatville to complement Chambers' Temple of Victory (no longer standing), this stone building with its Tuscan porticos contains iron plaques commemorating British military victories from Minden to Waterloo.

THE ICE HOUSE

Records say that the Ice House, in the heart of the Winter Garden, was in use in 1763. A door, of thick wood for insulation, opened up the north-facing entrance tunnel and the whole brick-lined structure was covered with earth. Winter ice was collected from Kew's original lake, and it took three days to store the 'harvest'. The ice - a great luxury - was packed in with straw and lasted long enough to be used in summer to cool drinks and keep food fresh.

THE CAMPANILE

One of the 39 listed buildings at Kew, the Campanile is in the Italianate Romanesque style of stock brick with red brick dressings. This classical 'bell tower' was designed by Decimus Burton as a disguised chimney for the Palm House boilers 100 m (328 ft) away. A tunnel for both the flue and a railway to carry coal linked the two buildings. Today, it carries hot water piped to the Palm House heaters from modern boilers near the Victoria Plaza.

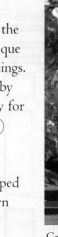

Campanile

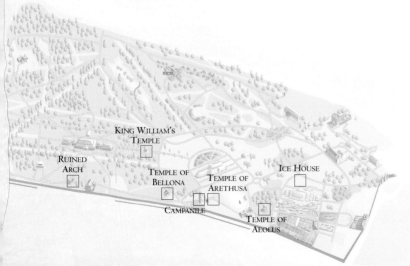

KEW'S GATES

Four of Kew's gates are Grade II listed - the Main Gate, designed by Decimus Burton in 1845 and completed the next year; the Lion Gate and Lodge, the Unicorn Gate, and the Victoria Gate, the latter three all being built during the second half of the 19th century.

Burton's Main Gate on Kew Green signified a change of attitude on the part of Kew's management, because Sir William Hooker, when he became Director, no longer required visitors to be escorted by gardeners. At the end of his first year, some 9,000 people passed through this grand entrance with its coat of arms and ornamental foliage.

The London and South Western Railway reached Richmond; another branch from Brentford to Willesden brought passengers to 'Kew Junction'; river steamers stopped at Kew for the Gardens; and by 1850, annual attendance had reached the heady heights of more than 150,000 visitors.

There is as much history behind Kew's gates as there is in the Gardens. They tell of expansion, of demands for public access and above all, of a fine monumental design tradition that is as impressive today as it was when they were built.

THE QUEEN'S BEASTS

These ten heraldic figures in front of the Palm House are Portland stone replicas of those which stood at Westminster Abbey during the coronation of H.M. Queen Elizabeth II. By the same sculptor, the late James Woodford, they were presented anonymously to the Gardens in 1956. The beasts, selected from the armorial bearings of many of the Queen's forbears, are judged to best illustrate her royal lineage. They are shown below as they appear with the Pond behind the onlooker.

The White Greyhound of Richmond

The Yale of Beaufort

The Red Dragon of Wales

The White Horse of Hanover

The Lion of England

THE PALM HOUSE POND

The Palm House Pond was part of the major transformation of Kew in 1845, under William Nesfield and Decimus Burton. Once a small remnant of Kew's original lake, it was enlarged and reshaped to provide a water feature in its own right, and a balance to the formal parterres. It also reflected the entire length of the Palm House in its waters, adding to the focal importance of the magnificent building.

A simple fountain was added in 1853 and the current sculpture of Hercules wrestling the river-god Achelous, was added in 1963. Based on an original sculpture by Francois Joseph Bosio (1814), which still sits in the Jardin de Tuilieres, Paris, this bronze cast was made by C. Crozatier for George IV in 1826, and once sat on the East Terrace of Windsor Castle. Two white marble Chinese Guardian Lions can be found on the Victoria Gate side of the Pond.

It is thought these date from the Ming Dynasty of 1368 – 1644 AD. They were presented to Kew by Sir John Ramsden in 1958.

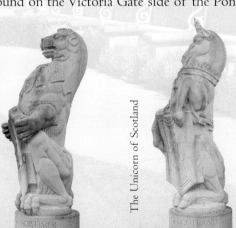

The White Lion of Mortimer

The Unicorn of Scotland

The Griffin of Edward III

The Black Bull of Clarence

The Falcon of the Plantagenets

Special places to see at Kew

PLANTS+PEOPLE, THE ORANGERY, THE NASH CONSERVATORY

When Sir William Hooker became Kew's first Director in 1841, he set about persuading the administrators of the need for a museum of economic botany, demonstrating the importance of plants to mankind.

In 1846, he displayed his own personal collection of specimens of textiles, gums, dyes and timbers in a fruit store and servants' quarters given to Kew by the Royal Family. Decimus Burton was then asked to adapt part of the building into the new Museum of Economic Botany, which opened on 20th September 1848 to instant success.

Collections grew with contributions from the Great Exhibition in 1851 and the Paris Exhibition of 1855. It soon became clear that the Museum was too small and Decimus Burton was asked to design a new Museum, twice the size, opposite the Palm House. Perversely, this new 1857 Museum was called Museum No 1, with the first renamed as Museum No 2 - now the School of Horticulture.

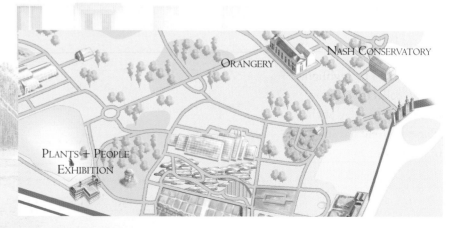

PLANTS+PEOPLE EXHIBITION

The study of plants used by people has formed an important part of the work of the Royal Botanic Gardens since the time of George III. In the 19th century, its aim was to aid the economic growth of the British Empire. Today, its focus is on the sustainable and medicinal uses of plants, and on conservation.

The Economic Botany Collections at Kew are the oldest and largest in the world and are of global significance. They consist of over 82,000 plant specimens, products made from plants, and tools used in their cultivation and processing. The highlights of the collection are on permanent display at the Plants+People Exhibition in Museum No 1.

THE ORANGERY

This 1761 building was designed by Sir William Chambers. It is the largest classical style building in the Gardens, measuring 28 m (92 ft) long by 10 m (33 ft) deep.

Princess Augusta's coat of arms was placed over the central bay of the facade in the 1840s, along with the Royal Arms and escutcheons with the monogram 'A' in honour of Queen Adelaide, wife of William IV.

Designed to hold orange trees, the light levels inside the building were too low to grow plants, even after glass doors were added at either end in 1842. Today, it is a deservedly popular, elegant café-restaurant.

THE NASH CONSERVATORY

Of a classical stone 'Greek temple' design, this is the oldest of Kew's 19th century glasshouses. It was originally one of four pavilions designed by John Nash for the gardens at Buckingham Palace, but William IV moved this one to Kew in 1836, after which it was adapted by Sir Jeffry Wyatville, the designer of King William's

Temple. All these associations mean it has major historical and architectural significance.

The Nash Conservatory first housed *Araucaria* and *Eucalyptus*; then Australian flora in 1854. In 1861, it became the Aroid House and over a century later it was used as a temporary home for large palms when the Palm House was restored in the early 1980s. The Nash Conservatory is now fully restored to take its rightful place among Kew's iconic buildings.

Art at Kew

Botanists have always needed artists. Plants die, dried specimens fade, but botanical art records plants in detail for posterity. Some botanical artists are professionals, accompanying collecting expeditions; some are gifted amateurs making their own collections of paintings and drawings for personal satisfaction.

THE MARIANNE NORTH GALLERY

Marianne North was a highly talented amateur with a great eye for detail. She learned to enjoy travel with her father, the MP for Hastings, whom she accompanied on his tours of Europe and the Middle East.

She was given lessons in flower painting by a Dutch woman artist and by Valentine Bartholomew, Flower-painter-in-Ordinary to Queen Victoria. On the death of her father, whom she described as "the one idol

and friend of my life", she started her travels at the age of 40 and, with several letters of introduction, began her search for the world's exotic flora.

Her travels were relentless, her productivity seemingly inexhaustible. Her first solo trip in 1871 saw her in Jamaica and North America; her next to Brazil; then Japan in 1875, returning home through Sarawak, Java and Ceylon. India inspired her to 200 paintings of buildings as well as plants.

She sketched rapidly in pen and ink on heavy paper, then the oils would come straight from the tube. Her palette was of bold and assertive colours and her enthusiasm evident, and just occasionally rather undisciplined.

She liked to paint plants "in their homes" in ecological settings with the addition of an occasional insect or other small creature, especially frogs; and sometimes against a mountain background.

In August 1879, Marianne North wrote to Sir Joseph Hooker, offering Kew not only her collection of paintings, but also a gallery in which to house them, her only stipulation being the use of a room as a studio. Kew, of course, accepted.

Her architect friend James Fergusson designed a gallery which mirrored her feelings for India, providing a verandah around the outside of the building; while satisfying his own ideas on lighting with large clerestory windows high above the paintings.

Miss North took charge of the hanging herself, arranging them in geographical order over a dado of 246 vertical strips of different timbers. The walls are virtually solid with paintings - there are 832 artworks all told, showing over 900 species of plants - a unique memorial to an equally unique woman.

Kew Gardens Gallery

Located in Cambridge Cottage, Kew Gardens Gallery has periodically-changing exhibitions of botanical art by both past and contemporary artists. This quiet and charming gallery is a favourite haunt for many of Kew's art-loving visitors. Because of the many facilities in Cambridge Cottage, it may be hired for private functions, such as weddings and other receptions. At these times, the Gallery is closed to the public.

From stone to canvas

Kew has held many major exhibitions over the past decade – from Chapungu in 2000, to Emily Young in 2003, and Chihuly in 2005/6. Kew has hosted the annual Garden Photographers' Association exhibition for several years, and new artistic events are constantly being planned for the Gardens, such as the Henry Moore exhibition in 2007/8. An exciting new gallery next to the Marianne North Gallery will open in 2009.

What's in a name?

THE SYSTEM AT KEW

Plants are known by a variety of names. They have both common names and scientific or Latin names. There's room for confusion as the same plant may be called by different common names. This is obvious with different countries and languages, but the same plant may have many regional names in the same country. For example, here in the UK, bird's-foot trefoil (*Lotus corniculatus*) is also known as hen and chickens, Tom Thumb, granny's toenails, cuckoo's stockings, and Dutchman's clogs. But any botanist, from Aberystwyth to Zagreb, knows what is meant by *Lotus corniculatus*.

Again, an English bluebell, like those in the woods at Kew, is *Hyacinthoides non-scripta*, but in Scotland, a bluebell is *Campanula rotundifolia*, a delicate little flower of dry grassy places. In England, however, *Campanula rotundifolia* is a harebell.

The advantage of the scientific name is that it's recognised everywhere. The advantage of a *Latin* scientific name is that as Latin is a dead language, it cannot change its meaning, but, at the same time, is very descriptive. There are no political or nationalistic overtones in Latin, either, so it is readily accepted throughout the world.

THE LINNEAN BREAKTHROUGH

The Swedish botanist Carolus Linnaeus (1707-78) is credited with the naming system used today for all living organisms - plants and animals. He was the first to group organisms into a logical hierarchy based on shared similarities. His plant classification was based on flower parts and remained unchallenged for about 200 years until the works of Charles Darwin and Gregor Mendel, with their respective theories of evolution and genetic inheritance, led to the desire to produce a *phylogenetic* classification, one which reflects evolutionary changes.

Nowadays, with DNA research undertaken at Kew, plants are also being grouped into their relationships according to differences in their genetic fingerprints (see p 86).

RECOGNISING PLANTS

This typical label shows the system used within the Royal Botanic Gardens, both at Kew and at Wakehurst Place. Here, it refers to the plant shown immediately on the right. Other plant labels may contain additional symbols or information.

Collector's or donor's code and plant's collection number

When shown, a Latin name here is the plant's family

Accession number: a unique number given to each plant or group of plants in Kew's collection

1992-123 HHME	CAMPANULACEAE
"PEACH-LEAVED BELLFLOWER" **Campanula persicifolia**	
F N	N. ASIA, EUROPE

When shown, N here states that the plant originated from natural (wild) source material

Plant's natural distribution range

Common name

F signifies that the identity of the plant has been fully verified by a botanist at Kew; P signifies part verification. (NB: on some labels • signifies full verification.) When shown, 'C' here denotes a commemorative tree

Scientific name shows genus and then species

PREDICTIVE TAXONOMY

The study of plant classification is called 'taxonomy' and it's carried out by taxonomists. Phylogenetic classification, based on relatedness, can be *predictive*. It happened with drugs from plants that may help with AIDS. An important chemical called castanospermine was found in an Australian tree in the genus *Castanospermum*. Taxonomists at Kew predicted that the chemical, or something very similar, might be found in the related Amazonian plant *Alexa* and were proved correct.

CLASSIFY OR DIE

There's a very basic human need to classify things. It probably developed in the hunter-gatherer days, when plants had to be classified into edible and poisonous, and that knowledge needed to be passed on in order to survive. Food, magic, medicine, fire, use as tools or weapons - different plants had different properties.

The ancient Greeks had words for plants. Theophrastus (d. 287 BC) classified, in a very basic way, the 500-odd plants in the

SOLANACEAE - ALL IN THE FAMILY

This family is wide-spread geographically and surprisingly diverse. It includes: potatoes, tomatoes, aubergines, capsicum and chilli peppers, ornamental solanums - *and deadly nightshade*, so do watch what goes into the ratatouille!

Athens botanic garden, while Dioscorides (c. 40-90 AD) wrote the first herbal, *Materia Medica*, which was used for about 1,000 years.

BASICS OF CLASSIFICATION

With plants, the basic group is a *species*, with a unique combination of leaf, stem, flower, fruit and seed characteristics. Species are often found in a particular geographical area and don't usually interbreed. When species have general characteristics in common, they are themselves grouped into a *genus* (plural *genera*). Several genera with basics in common can be put into a larger group, called a *family*, such as 'grasses' or 'bellflowers'.

The Order Beds (see p 51) are the ideal place to learn more about taxonomy because here, members of the same family are planted together so their similarities and differences can be seen very easily.

LEARNING LATIN FROM PLANTS

Plants have first a genus name, like a surname, then a species name, which is like a given name. The two names identify the plant and often describe it entertainingly, the Latin giving excellent clues as to its derivation. For example:-

GIANT HORSETAIL

Equisetum giganteum, coming from *Equus*, horse and *setum*, bristle, together with *giganteum*, giant.

DAISY

Bellis perennis, coming from *bellus*, pretty and *perennis*, perennial.

STINKING IRIS

Iris foetidissima, coming from *iris*, a rainbow for the many colours of the flower and *foetid*, smelly with *-issima*, most - the smelliest!

Royal Kew

KEW PALACE

After a ten year conservation project with the help of the Heritage Lottery Fund, Kew Palace has been sumptuously restored and is now open to the public.

Built in 1631 for the rich merchant Samuel Fortrey (whose initials are on the front decoration of the palace) this is the oldest building on the Kew site. Originally known as the Dutch House, it was leased by the Royal Family in 1728 and became home to the three elder daughters of George II. In 1731, George II's son Frederick, Prince of Wales, leased a house opposite Kew Palace, known as Kew Park, and converted it into a large Palladian palace known as the White House. It was here that George III grew up and lived until 1802, when he purchased Kew Palace and demolished the White House. It was at Kew Palace that his wife Charlotte gave birth to George IV, and here that they found retreat during the King's bouts of 'madness'.

After Queen Charlotte died in 1818, Kew Palace was closed. In December 1896, Queen Victoria agreed to Kew's acquisition of the Palace, providing there was no alteration to the room in which Queen

Charlotte died. In 1898, the Palace passed to the Department of Works and opened to the public for the first time. Today the Palace is in the trust of Historic Royal Palaces.

Kew Palace was an intimate and domestic residence for the royals. It has recently been restored to its condition in 1804. Its interiors provide a stark contrast to the typical expectations of royal palaces and it reveals a fascinating insight into Georgian taste and style. Previously closed areas of the Palace are open to the public, including

rooms on the second floor that have remained untouched and unseen since the Royal Family's departure nearly 200 years ago. Pertinent artefacts that belonged to the King and his family tell of their own personal history. Many areas have been left exposed to reveal the internal construction of the Palace and the painstaking conservation process.

For more details and the story of the restoration go to **www.hrp.org**

The charming Queen's Garden behind the Palace is described in more detail on p 52.

Queen Charlotte's Cottage

The origins of this cottage may well have been the single storey building provided for the Menagerie keeper. There is no doubt that Queen Charlotte was given the building in 1761 when she married George III and that she extended the property upwards by a floor and also in length.

The picturesque house in its 'cottage ornée' style was used by the family as a shelter, and for snacks and occasional meals. The large ground floor room had Hogarth prints on the walls, removed in the 1890s but replaced in 1978. A curved staircase leads to the picnic room, with painted flowers climbing the walls and bamboo motif pelmets and door frames. These charming murals were painted by Princess Elizabeth herself, which emphasises the family's feeling for the cottage.

The cottage remained private until 1898, when Queen Victoria ceded it and its 15 hectares (37 acres) to Kew to commemorate her Diamond Jubilee. The grounds had rarely been visited; trees were lying where they had fallen, and one condition the Queen made was that the grounds should be kept in their naturalistic state. This condition was supported by the Linnaean Society on behalf of all ornithologists to maintain the area as a suburban haven for birds. That is how today's Conservation Area first came into being (see pp 62 & 63) and how one of London's finest bluebell woods is kept intact.

*Queen Charlotte's Cottage is maintained and administered separately from Kew by Historic Royal Palaces and opening times are limited. See **www.hrp.org** for more details.*

How Kew Grew

For Kew's heritage year festival in 2006, research into the original locations of Augusta's follies at Kew, and Queen Caroline's designs in the neighbouring Richmond estate, was carried out on an exciting new level. The result is a 12 minute computer animation film explaining the history of Kew as it had never been seen before. 'How Kew Grew', made with the expert help of King's Visualisation Lab, at King's College London, is an ongoing project that hopes to expand by 2009, and continue to reveal the layers of the history of Kew's landscape and buildings. 'How Kew Grew' is available on DVD at the Victoria Plaza.

1763

The history of Kew

Princess Augusta

THE ESTATES

The very early history of Kew Gardens from 1718 is touched on in the story of Kew Palace (see previous pages), and taken up here with the marriage of King George II's son, Frederick, Prince of Wales, to Princess Augusta of Saxe-Gotha in 1736, when they lived in the White House, near what is now Kew Palace.

Frederick and Augusta were garden enthusiasts, and were helped by the Earl of Bute, who advised them on obtaining plants and landscaping and then became the 'finishing tutor' to the Wales's son, later George III.

GRAND PLANS

The Prince's garden increased in both content and ambition. With his dilettante interest in art, literature and science, Frederick's 1750 plans for the garden embraced trees, exotics, and a wish for an aqueduct and a "mound to be adorned with the statues or busts of all these philosophers and to represent the Mount of Parnassus."

Before these plans could be realised, he died, after a bout of pleurisy and from a burst abscess in his chest, possibly caused by a blow from a cricket ball some considerable time earlier. His death was lamented by England's gardening fraternity and the botanist Dr John Mitchell declared that "Planting and Botany in England would be the poorer for his Passing."

PRINCESS AUGUSTA AND GEORGE III

In 1752, Princess Augusta instructed her head gardener, John Dillman, to complete the works planned by her late husband. With the very able help of the Earl of Bute, the development of Kew as a serious botanic garden was well under way, driven by Bute's desire to have a garden which would "… contain all the plants known on Earth." It was on nine acres of ground next to the Orangery that the first serious botanical collection at Kew was planted. Princess Augusta was, in effect, the founder of the botanic gardens at Kew.

Bute introduced the Princess to William Chambers, an ambitious young architect who had previously submitted plans for a projected mausoleum for her dead husband, and in 1757, Chambers was appointed tutor in architecture to the future George III, who came to the throne in 1760. See p 67 for more on Chambers and p 70 for his decorative buildings.

In 1764, Capability Brown, now Master Gardener at Hampton Court, was instructed to redesign the neighbouring Richmond estate. In creating his 'natural' landscapes, Brown removed most of the decorative buildings, 18 houses in West Sheen and the whole of the riverside terrace. He was almost universally disliked for this at the time, but 20 years later, the mature gardens were admired, typifying the clash of styles; the formal decorative against the 'new' naturalistic.

With his succession, George III inherited the Kew and Richmond estates and joined them in 1802 with the closure of Love

Lane and the removal of its boundary walls. Then with the support of Sir Joseph Banks, the Gardens took on a new phase of plant collecting.

Sir Joseph Banks (1743-1820)

A wealthy entrepreneur and natural history enthusiast, Banks went on several collecting expeditions including, between 1768-71, James Cook's round the world expedition in the Endeavour. He paid for his own passage and those of eight companions,

including botanists, artists and a secretary. He was elected President of the Royal Society in 1778 and held the post for 41 years. He used his influence, taking on, in his words, "a kind of superintendence" to promote the Gardens at Kew and without his guidance, it is doubtful if Kew would have evolved into the internationally respected institution it is today. He resolved that new plants should always be seen first at Kew and organised collecting expeditions throughout the world. He also asked diplomats, army and navy officers, merchants and missionaries to remember Kew on their travels.

Under Banks' directions, Francis Masson went to South Africa and North America; Robert Brown and Peter Good journeyed to Australia and William Kerr started collecting in China. Two more, Allen

Cunningham and James Bowie, spent two highly productive years in Brazil collecting bromeliads and orchids.

Banks' last visit to Kew

In 1819, Banks visited Kew for the last time, specifically to see a plant that Francis Masson had introduced in 1775; the African cycad *Encephalartos altensteinii* which had just produced its very first cone, 44 years after arriving. Sir Joseph Banks died on 19 June 1820.

The next forty years saw the Prince Regent become George IV, to be succeeded by William IV and then, in 1837, Queen Victoria. After a period of decline from 1820, when Banks died, Kew was given to the nation in 1840, when it was transferred to the Office of Woods and Forests.

Mutiny on the Bounty

Breadfruit was introduced to the West Indies from its native Tahiti to feed plantation slaves. Sir Joseph Banks chose Captain Bligh to sail and collect breadfruit seedlings in 1789. The mutiny on his ship, HMS Bounty, delayed the breadfruit's arrival until 1792. What's more, the slaves preferred plantains.

Wardian cases

Sealed glass cases named after their inventor, Nathaniel Ward, were absolutely key to the transfer of living plants from the furthest corners of the world back to Kew. They were used first in 1835 and last in 1962 to bring plants from Fiji.

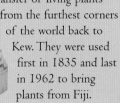

The history of Kew

FROM STRENGTH TO STRENGTH

When Sir William Hooker took charge of Kew in 1841, collecting started again in earnest and a new era dawned for the Gardens. With Decimus Burton's new buildings, ailing plants were revived and room was made for more; and with the landscape design skills of W A Nesfield, trees and shrubs were planted in family groups throughout the site.

Joseph Hooker, Sir William's son, was an avid explorer and collector who later took over as Director from his father. In 1853, some 4,500 herbaceous plants were in cultivation. By 1864, there were more than 13,000 species, with over 3,000 species of trees and shrubs in the Arboretum.

Today's living collections at Kew are scientific resources for plant conservation and are used for research by botanists in the Herbarium and the Jodrell Laboratory (see pp 85 - 88). The palm collection is particularly fine as a result of botanical expeditions and similar expeditions are still sent out from Kew to bring back specimens collected under guidelines laid down by the Convention on Biological Diversity.

THE 20TH CENTURY

The Gardens have been interrupted in their growth from time to time. Suffragettes burnt down the Refreshment Pavilion in 1913. There was a violent storm in 1916, a drought in 1921, an exceptionally heavy snowfall in 1926 and the Great Storm of 1987 did considerable damage.

But Kew has gone from strength to strength, and among the many high spots have been the restorations of the Palm and Temperate Houses, and the building of the Princess of Wales Conservatory and the Davies Alpine House.

ST ANNE'S CHURCH, ON KEW GREEN

Consecrated in 1714, this handsome church has enjoyed royal patronage for many years. It was enlarged by George III in 1770 and by William IV in 1837. It houses tombs and memorials to many of Kew's past administrators, botanists, artists and gardeners and the centuries-old relationship between St Anne's Church and Kew Gardens is very much alive today.

FAMOUS FLOWERINGS

The Titan arum, *Amorphophallus titanum*, shown right, flowered in June 1889, for the first time ever outside its native Sumatra. Since June 2001, its annual flowerings at Kew have attracted large crowds, but visitors never stay long, due to its truly disgusting odour.

84

The role of Kew today

Kew was founded in a time of great curiosity about the world. The Georgians collected plants from distant countries that were being explored for the first time. During Victorian times this curiosity was intensified at Kew and put onto a more scientific footing. In 1984 the gardens were made autonomous from government under a new Board of Trustees. Collection and research have continued into the 21st Century, making Kew the international resource on plant identification and knowledge.

Recent years have seen a growing recognition of conservation issues, and Kew's work is increasingly valued for its role in habitat conservation and measures to counter climate change. The vital role that plants play in the ecosystem is not lost on world leaders. The Convention on Biological Diversity, which Britain ratified in 1994, requires signatories actively to promote understanding and awareness of biological diversity and the need for conservation.

Kew plays a significant role in this 'active promotion' as demonstrated in the mission statement.

Kew's Mission

To inspire and deliver science-based plant conservation worldwide, enhancing the quality of life.

This will be achieved by the following objectives, allowing Kew to maintain impact and excellence in science-based plant and fungal conservation.

- Engagement and learning - engaging large and diverse audiences in inspirational, enjoyable experiences, which encourage learning and positively change attitudes to plants and fungi and their conservation.

- Global conservation and sustainability - using Kew's unique combination of resources to support and deliver conservation and sustainability projects around the world.

- High quality, high impact research - discovering and sharing information about plants and fungi, using and enhancing Kew's globally important scientific collections and knowledge.

- World-class horticulture - developing, promoting, teaching and delivering best practice in horticulture.

- Capacity building and collaboration - achieving plant conservation through capacity building, partnership and collaboration, higher education and training in local, national and international frameworks.

- World heritage - caring for and enhancing Kew's heritage collections, buildings and landscapes, and communicating their universal value.

- Policy and advocacy - providing sound, scientific evidence and advice to advance and support informed debate, policy-making and action.

Kew today

RESEARCH AND CONSERVATION

WHY RESEARCH?

By studying plant structures, genetic material and biochemistry, relationships between plants can be clarified, leading to many practical applications. At Kew, plant anatomy, cytogenetics and other laboratory-based research is carried out in the Jodrell Laboratory. Located by the Aquatic Garden, it is not open to the public, but it is possible to see the laboratory in action from the outside.

In the Herbarium, plants are identified, named and classified, resulting in detailed studies of particular groups of plants - how they interrelate, and how they differ from each other. Carrying out surveys of vegetation in many different parts of the world is the very foundation of other plant research or conservation projects. The Herbarium, not open to the public, attracts an average of 50 researchers from around the world every week.

GENETIC 'FINGERPRINTING'

Kew is at the cutting edge of research. Scientists in the Jodrell Laboratory have pioneered techniques using the chemicals that plants naturally produce to reveal relationships between them.

Another technique involves treating plant tissues by 'painting' their chromosomes with special dyes which glow under a microscope's light. Different colours identify the unique chromosomes from different plants and the parents of a hybrid can be traced this way. Genetic fingerprinting makes use of the built-in instructions each plant has for making chlorophyll. Each plant species has its own set of instructions and when different sets are compared, close matches reveal close relationships. Results often confirm current thinking, but occasionally previously unknown relationships have been revealed.

A VITAL ROLE, BEAUTIFULLY PACKAGED

Through research, Kew's scientists are constantly taking enormously important steps towards harnessing the abilities of plants not only to feed the world more effectively, but to help fight disease and other problems.

Kew's role, then, is predominantly in gaining and distributing knowledge of the plant kingdom and its relationship to mankind. That this globally vital role is so beautifully and interestingly packaged at Kew and at Wakehurst Place, is a bonus for all of us.

Plant Collections

As the world faces a growing environmental crisis, the work of botanic gardens becomes more deeply involved in conservation and the sustainable use of ecosystems.

Living collections of plants have become a vital part of botanical research and conservation. At Kew, the collection of 31,000 taxa, or groups of plants, contains 15 species that are totally extinct in the wild, and 2,000 more included in lists of threatened and endangered species. Kew's work on micropropagation - multiplying plants by reproducing them from buds or tiny seeds - leading to eventual reintroduction in the wild, is critically important to biodiversity.

Kew holds more than living plants, as can be realised from the great numbers of carefully preserved specimens, both old and recently collected, in the Herbarium. There are currently some 7 million specimens in the Herbarium, representing 98% of all the known genera (groups of similar species) in the world. The mycological specimen collection of 800,000 mushrooms, toadstools and other fungi, is one of the largest and most important in the world.

Economic Botany

Kew's Economic Botany Collection was started by Sir William Hooker, Kew's first Director, over 150 years ago (see pp 74). The 82,000 items include one of the finest wood collections in the world (32,000 samples from 12,000 species), thousands of bottles of oils, and thousands of other plant-based artefacts and oddities. Many items from the collection are on view in the Plants+People Exhibition in Museum No 1.

Conservation and sustainable use are the watchwords behind Kew's work today. Kew's Millennium Seed Bank, at Wakehurst Place (see p 91) is a perfect example of our crucially important conservation work. Kew acts as adviser to the UK government on the Convention on International Trade in Endangered Species of Flora and Fauna (CITES) and also on international conservation legislation. Conservation genetics is important work at Kew, looking at the genetic diversity of populations of endangered plants, to avoid interbreeding and so make sure the populations remain as healthy as possible.

Kew's work, then, as in its early days, still embraces classification of the world's plants and the promotion of economic botany. Today, though, the emphasis is not on economic botany for the growth of 'Empire', but more for the benefit of mankind as a whole.

Extinct British grass grows at Kew

Extinct in the wild, having been seen last in 1963, *Bromus interruptus* is, for botanists, the UK's most famous grass. It is being grown in Kew's living collections and is a subject of a Biodiversity Action Plan led by Kew. Vast numbers of plants have been preserved in the Herbarium on Herbarium Sheets. A well-collected herbarium specimen comprises the vital parts of the plant - flowers and fruits - as well as some leaves attached to the stem. The label - essential to the collection - carries field notes, the collector's name, a serial number and other important details. The Herbarium Sheet for *Bromus interruptus* is shown here.

Kew today

LEARNING

It is one of Kew's stated aims to increase people's understanding of the plant kingdom and so learning is seen as playing an absolutely key role. So much so, that 'learning' at Kew is divided into three: Higher Education and Training; Public Education and Interpretation; and Schools Education, in order to meet the needs of different audiences.

PUTTING PLANTS ON THE CURRICULUM

Botany and conservation are very much on today's curriculum. Kew's mission in education is to make the plant kingdom interesting, fun - and important. Caught young, children are conservationists for life.

More than 90,000 schoolchildren visit Kew and Wakehurst in booked groups each year, armed with work sheets and education packs, or enjoying even greater rewards from a tour guided by one of Kew's teaching team. Museum No I, where the Plants+People Exhibition regularly astounds visitors of all ages, let alone schoolchildren, has a popular classroom facility.

PLANTING IDEAS

The Outreach Education Programme sees Kew visiting schools, running anything from week-long specially tailored workshops in a mobile classroom, to a single talk loaded with specimens and demonstrations and long remembered through the posters and leaflets left behind.

At Kew's School of Horticulture, students can gain the three-year Kew Diploma of Horticulture, and prestigious research degrees can be gained here, too. There is an expanding programme of international courses in herbarium techniques and botanic garden management.

CLIMBERS AND CREEPERS

In the summer of 2004, Kew proved its determination to make serious botany great fun for children with 'Climbers and Creepers', Britain's first interactive botanical play zone. A riotous success from the outset, with over 120,000 young visitors in its first year, Climbers and Creepers engages children from around 3-9 years in the pleasures of learning more about plants and their relationships with animals and people.

Children can climb inside a plant to learn about pollination, and see the dangers insects face when they are 'eaten' by a giant pitcher plant. Climbers and Creepers is open year-round for families to drop in casually and for school parties to visit on a more organised basis.

Learn more at www.kew.org/climbersandcreepers.

Please note: while properly-trained and security-cleared Kew staff and volunteers work in the zone, parents remain fully responsible for their children, and are advised to keep an eye on them at all times.

How Kew is funded

KEW'S WORK BENEFITS THE WHOLE WORLD, AND YOU CAN BE PART OF IT.

"Today the Royal Botanic Gardens, Kew, is increasingly involved in the conservation of our environment as thousands of species of plants face extinction. And the simple fact is that without plants, human life on this earth will cease to exist.

Kew's botanists and horticulturists are able to solve global problems by advising on environmental management, conserving endangered species, and searching for alternative crops as sources of food, animal fodder, fuel and medicine.

In supporting the Friends, you will be helping to finance the vital work Kew is doing to help save plant life and the world.

Thank you for your support."

The Royal Botanic Gardens, Kew, is a registered charity, set up in April 1984 under the terms of the National Heritage Act, 1983. Responsibility for the Gardens then passed from the Ministry of Agriculture, Fisheries and Food (MAFF) to a Board of Trustees.

Kew's functions are broadly defined in an Act of Parliament and are today firmly established in the Mission Statement (see p 85). The Gardens are in part funded by Defra (the Department for Environment, Food and Rural Affairs).

KEW FOUNDATION

In March 1990, Kew Foundation was set up with the sole aim of raising funds for projects not covered by grant aid and self-generated money. The Foundation has proved an unqualified success, raising in excess of £2 million a year towards Kew projects.

FRIENDS OF KEW

June 1990 saw the launch of 'Friends of Kew', with membership bringing special privileges and opportunities to those wishing to know more about Kew specifically and horticulture generally. Membership income and that generated by fees for various events all contribute much-needed revenue. There's no better way of supporting Kew's mission than to become a Friend. Here are some of the benefits:

- unlimited free entry to Kew Gardens and Wakehurst Place
- six free day passes for family members
- free entry to 15 other gardens nationwide
- a free copy of Kew magazine - sent to you four times a year
- exclusive events for Friends (walks, talks and behind-the-scene activities)
- privilege discounts at Kew and Wakehurst Place shops, and on Kew Adult Education courses ranging from gardening to botanical art
- advance notice of our many events and priority booking for 'Summer Swing' concerts

Details on how to join Friends of Kew are in the leaflets available at the Victoria Plaza and on the entry maps handed to visitors. Or call 020 8332 3200 or visit www.kew.org.

RBG KEW ENTERPRISES

To supports its important work in the UK and abroad, Kew generates income from a variety of sources. These range from gate receipts to commercial activities such as catering, retail and licensing. Kew Enterprises also runs popular public events in the Gardens, and hires out its iconic venues.

Wakehurst Place and the

The Royal Botanic Gardens leased Wakehurst Place from the National Trust in 1965. Set in the beautiful High Weald of Sussex, the elegant sandstone Elizabethan manor house is surrounded by a mixture of formal gardens and woodland walks.

Wakehurst Place enjoys a mild climate, a high rainfall and moisture-retentive soils, allowing many important groups of plants, difficult to grow successfully at Kew, to flourish. It is home to an outstanding collection of temperate flora from eastern Asia, South America, Australia and New Zealand built by Gerald Loder between 1903 and 1936. On Loder's death, Wakehurst Place was bought by Sir Henry Price and in his care the estate matured richly. Sir Henry left Wakehurst Place, with a sizeable endowment, to the nation in 1963.

Today, visitors can walk through realistic forest environments, such as New England's 'fall colour' or the majestic giant redwoods of California. With extensive water gardens, a steep Himalayan Glade, charming walled gardens nearer the mansion and excellent facilities for visitors, Wakehurst Place is a deservedly popular attraction.

But, with Kew's mission for conservation and serious botanical research, it is only to be expected that there is more to Wakehurst Place. And there is.

LODER VALLEY NATURE RESERVE

The Loder Valley Nature Reserve encompasses three major types of habitat; woodland, meadowland and wetland. It is living conservation and restoration at its best, since it preserves not only different habitats complete with their own birds, mammals and invertebrates, but also traditional countryside management and its resultant crafts and products. Coppicing, for example, not only produces wood for Bar-B-Kew charcoal and country products such as hurdles, besoms and rustic furniture, but also creates a habitat that allows the reintroduction of the native dormouse. A new bird hide opened in late 2006 within the Reserve to allow greater enjoyment of the wildlife here.

THE FRANCIS ROSE RESERVE

The Francis Rose Reserve is probably the first nature reserve in Europe to be dedicated to mosses, liverworts, lichens and filmy ferns (cryptogams). Francis Rose was a renowned botanist who pioneered the study of these plants and the sandrock outcrops of the Sussex High Weald where they have taken refuge.

The reserve stretches from the Himalayan Glade in the public part of the gardens, to Tilgate Wood, just inside the Loder Valley Nature Reserve. In this stretch lie some of the UK's best sandstone rock habitats, already designated a Site of Special Scientific Interest. Information panels along the sandrock outcrops highlight the species to be seen and the habitats they enjoy, as well as the environmental threats they face.

A dormouse at Wakehurst Place - part of English Nature's Species Recovery Programme

Millennium Seed Bank

THE MILLENNIUM SEED BANK

This magnificent £80 million project was part funded by the Millennium Commission and supported by the Wellcome Trust, one of the world's premier medical charities, and Orange plc, together with many other organisations and individuals. Its aim is to conserve biodiversity by storing the seeds of not only every native plant in Britain, but even more importantly, protecting 24,000 other species against extinction around the globe. It is probably the most ambitious conservation project in the world and its interactive exhibition is a magnet for anyone wishing to learn more about plants and their intimate relationship to mankind.

Seeds have an astonishing ability to cling on to life against incredible odds. They'll certainly survive the exceptionally cold and dry conditions in the Millennium Seed Bank vault. Visitors can learn:-

- Why Kew started this international project and its importance to future generations
- How seeds are collected in the wild
- How the laboratories and seed preparation areas go about their work
- How the vaults will safeguard 24,000 plant species for hundreds of years.

The Royal Botanic Gardens, Kew would also like to thank both the National Trust and Defra, without whose support this idea would not have turned into the splendid and vitally important reality it has become.

The Wellcome Trust Millennium Building, named after the Wellcome Trust, who have donated over £9 million in recognition of the importance of plants as sources of medical benefits.

LOCATION

Open from 10 a.m. every day of the year except Christmas Eve and Christmas Day, Wakehurst Place is around 15 minutes from Junction 10 on the M23, from where it is clearly signposted to just north of Ardingly on the B2028. Haywards Heath station is six miles away by taxi and the estate is served by a regular bus service - for current details visit the Metrobus website on www.metrobus.co.uk or call 01293 449 191.

For the latest information about visiting Wakehurst Place including opening times and admission prices, please visit the website at www.kew.org, email wakehurst@kew.org or telephone 01444 894066 (24 hours).

General information

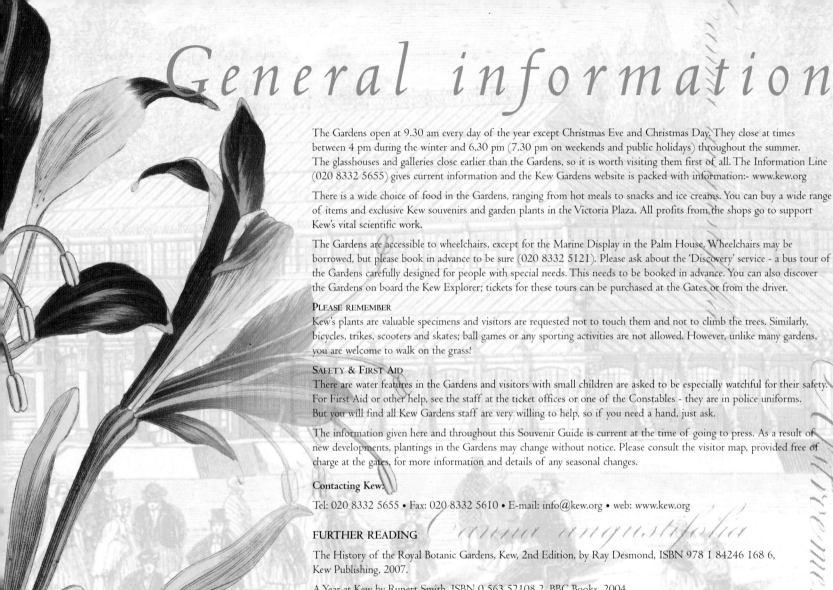

The Gardens open at 9.30 am every day of the year except Christmas Eve and Christmas Day. They close at times between 4 pm during the winter and 6.30 pm (7.30 pm on weekends and public holidays) throughout the summer. The glasshouses and galleries close earlier than the Gardens, so it is worth visiting them first of all. The Information Line (020 8332 5655) gives current information and the Kew Gardens website is packed with information:- www.kew.org

There is a wide choice of food in the Gardens, ranging from hot meals to snacks and ice creams. You can buy a wide range of items and exclusive Kew souvenirs and garden plants in the Victoria Plaza. All profits from the shops go to support Kew's vital scientific work.

The Gardens are accessible to wheelchairs, except for the Marine Display in the Palm House. Wheelchairs may be borrowed, but please book in advance to be sure (020 8332 5121). Please ask about the 'Discovery' service - a bus tour of the Gardens carefully designed for people with special needs. This needs to be booked in advance. You can also discover the Gardens on board the Kew Explorer; tickets for these tours can be purchased at the Gates or from the driver.

PLEASE REMEMBER

Kew's plants are valuable specimens and visitors are requested not to touch them and not to climb the trees. Similarly, bicycles, trikes, scooters and skates; ball games or any sporting activities are not allowed. However, unlike many gardens, you are welcome to walk on the grass!

SAFETY & FIRST AID

There are water features in the Gardens and visitors with small children are asked to be especially watchful for their safety. For First Aid or other help, see the staff at the ticket offices or one of the Constables - they are in police uniforms. But you will find all Kew Gardens staff are very willing to help, so if you need a hand, just ask.

The information given here and throughout this Souvenir Guide is current at the time of going to press. As a result of new developments, plantings in the Gardens may change without notice. Please consult the visitor map, provided free of charge at the gates, for more information and details of any seasonal changes.

Contacting Kew:

Tel: 020 8332 5655 • Fax: 020 8332 5610 • E-mail: info@kew.org • web: www.kew.org

FURTHER READING

The History of the Royal Botanic Gardens, Kew, 2nd Edition, by Ray Desmond, ISBN 978 1 84246 168 6, Kew Publishing, 2007.

A Year at Kew by Rupert Smith, ISBN 0 563 52108 2, BBC Books, 2004.

A Vision of Eden: The Life and Work of Marianne North, ISBN 1 84246 049 8, Kew Publishing, 2002.

The World of Kew by Carolyn Fry, ISBN 978 0 56349 378 5, BBC Books, 2006.

Kew Palace: The Official Illustrated History by Susanne Groom and Lee Prosser, ISBN 978 1 85894 323 7, Merrell Publishers Ltd, 2006.

Palaces and Parks of Richmond and Kew volume II, by John Cloake, ISBN 978 1 86077 023 4, Phillimore and Co Ltd, 1996.